Agile Principle Raw & Uncut

Agile Fundamentals

PHILL AKINWALE, PMP, PMI-ACP, PSM, CSM

ROY SCHILLING, PMI-ACP, CSP, PSM, CSM

Praizion Media

Real World Project Management Training Solutions

Agile Principle Raw and Uncut published by Praizion Media (V8)
P.O Box 22241, Mesa, AZ 85277
E-mail: info@praizion.com
www.praizion.com

Authors
Phillip Akinwale, MSc, OPM3, PMP, PMI-ACP, CSM, PSM, PSPO, PAL, SPS
Roy Schilling, CSP-SM, CSP-PO, ACP, ICP, ICP-ACC, PSM

ISBN 978-1-934579-77-0

9 781934 579770

Printed in the United States of America

Table of Contents

Table of Figures

WHY AGILE?

"It's not the strongest of the species that survives, nor the most intelligent, but the one most adaptable to change"
Leon C Megginson

Agile has become mainstream! It is everywhere. Let's discuss the rationale behind adopting Agile thinking, practices and methods. Many organizations are seeking newer approaches due to changes in technology, industries, and markets. Long-running projects are no longer feasible, there is a need for speed to stay ahead of competitors and meet customer expectations. This need for agility forms the primary motivation for embracing Agile.

For those preparing for the CAPM, PMP, PMI-ACP, or even the PSM exams, mastering the relevant content is crucial. This book is designed not only to introduce readers to the foundational principles and ideologies of agile but also to guide them towards success in various PMI exams and the PSM. In terms of depth and complexity, the PMI-ACP exam goes deeper into advanced agile topics, the PMP covers it to a lesser extent, the CAPM offers an introductory perspective, and the PSM focuses on Scrum. It's noteworthy that the PMI-ACP exam tends to emphasize situational and people-centric questions more than purely technical ones.

Initially, organizations often express their desire to adopt Agile to achieve speed, efficiency, and cost-effectiveness. While these are valid objectives, the true goal is continuous improvement and growth. Agile enables organizations to facilitate change more rapidly and deliver higher-quality outcomes. It fosters evolution by identifying and implementing better processes and product development approaches that ultimately satisfy customers. It is crucial to understand that Agile is should not be treated as a dogmatic belief system, but rather as a means to achieve what makes the most sense in each unique context. The emphasis should be on protecting the organization and serving the customers.

CHAPTER

1

AGILE BASICS

AGILE OR PREDICTIVE?

Let's discuss which delivery method is best. There is a diagram that will facilitate this decision. It is called the Stacey complexity model, and this is just one mechanism of determining what methods you really should be using.

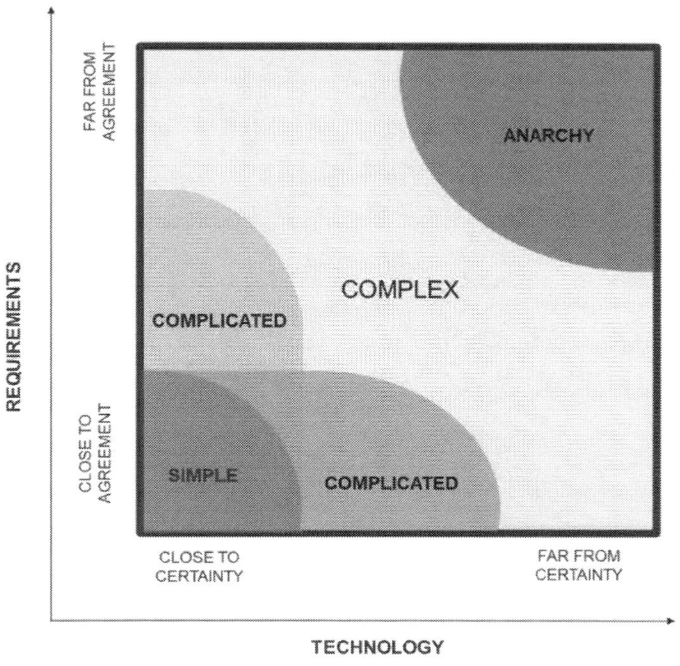

Figure 1: The Stacey Complexity Model

Consider the complexity model, and let's start from ***Close to Certainty*** in the lower left-hand corner down here in this ***SIMPLE*** space. It is where the traditional project management works exceedingly well. It is where things are very well understood. This refers to requirements that are close to certainty, and we also know that these are the requirements that we definitely want to implement.

Not everything works this way, and when you think about predictive project management (the traditional approach), its history came from construction, so that's where things were had to be very well-known, very well-understood because you're dealing with physical products. And I understand in a lot of your situations in your work; you may also be dealing with physical products and we're going to be talking about that in respect to Agile as well. But when you're dealing with uncertainty, when you're dealing with things where you have to understand, you have to experiment and learn more about what the solution should be or what the problem is, then that's where Agile starts to take hold. That's where Agile fits and finds its niche.

In complex situations where requirements and circumstances are uncertain, Agile methodologies thrive by fostering experimentation. Unlike the traditional predictive approach, which follows sequential phases of design, analysis, and building, Agile promotes a more iterative and adaptive process. Agile encourages trying out ideas, conducting small-scale design and analysis, and using the results as valuable feedback for subsequent experiments.

While the term "experiment" is frequently used in Agile, it does not imply a consistent expectation of failure. Rather, Agile encourages conducting survivable experiments—approaching tasks with the assumption that the chosen approach is correct but remaining open to adjustments and improvements along the way. This flexibility and adaptability are at the core of Agile's power, and you might already be implementing some Agile practices without realizing it. It's important to recognize Agile as a tool that can be tailored to suit your specific context and needs, without being bound to rigid rules or methodologies.

To successfully implement Agile, several key elements contribute to its effectiveness:

1. Buy-in from the organization: It is crucial for senior leadership and stakeholders to understand and accept Agile as a different way of working. This includes acknowledging the need for revised approaches to monitoring progress, reporting status, and

supporting teams. Organizational culture often presents the biggest obstacle to Agile adoption, which we'll discuss further.

2. T-shaped skills: In addition to deep subject-matter expertise, Agile encourages individuals to broaden their skill sets and collaborate across disciplines. Stepping outside one's comfort zone and assisting others in their areas of expertise not only promotes personal growth and career development but also enhances team dynamics and learning.
3. Risk reduction and resource allocation: Agile shifts the focus from resource allocation to team collaboration and collective responsibility. By assigning work to teams and empowering them to figure out how to accomplish it, organizations can reduce risk and improve overall efficiency.
4. Psychological co-location: While physical co-location may not always be feasible, especially in remote work settings or geographically dispersed teams, fostering a sense of psychological co-location is essential. Leveraging communication and collaboration tools, such as video conferencing and collaborative platforms, helps teams stay connected, aligned, and engaged. Socialization and team building activities should also be intentionally incorporated into the Agile process.

At Praizion, we have extensive experience working with remote teams facing similar challenges, and we understand the importance of social interaction and teamwork in Agile environments. Building a great team involves not only effective collaboration but also intentional socialization, and we encourage integrating both aspects into the Agile process.

The Continuum of Life Cycles

Building on the Stacey model, the next image is called the PMI Continuum of Life Cycles. Now, you could be anywhere on a continuum between predictive and agile.

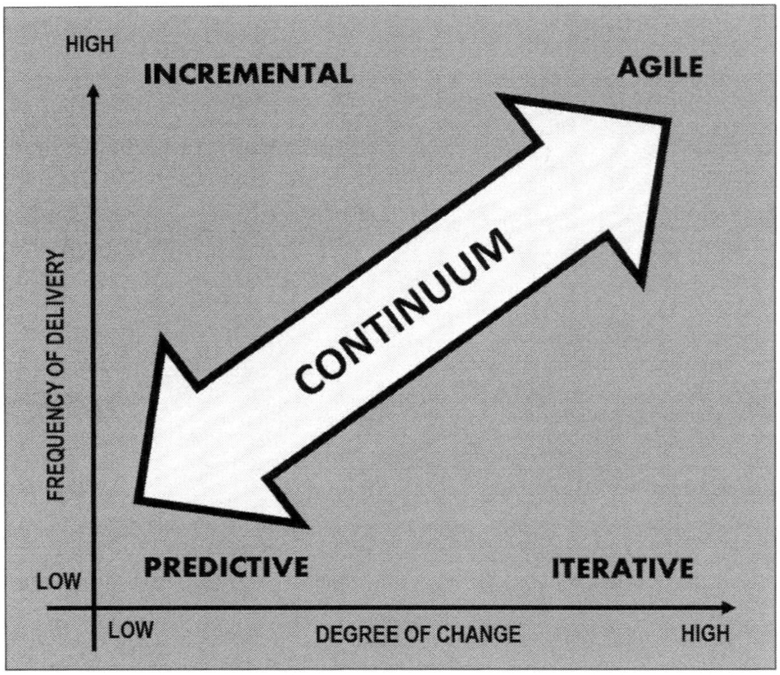

Figure 2: The Continuum of Life Cycles

Looking at this chart from the PMI's Agile Practice Guide, we can see a continuum of life cycles. And what this espouses is that you could be anywhere on this continuum between Predictive and Agile. There are some instances where you may need to develop and deliver in an incremental fashion and those where you need to deliver in an iterative fashion.

For long we've heard Predictive and Agile messages are at war with each other. And while they're not at war with each other, it would be foolish to postulate that the same practices, and a dare say, principles can be used in both sides of this continuum without any modifications. On the Predictive side of the continuum, you will find certain principles, ideas and methods more prevalent than in the Agile side of the continuum.

I call this "a new project management idea" because I want you to think about plan-driven projects in a non-threatening way to Agile projects in vice versa.

Every project no matter where you are is going to be on a continuum. It's either going to be very predictive or very agile or somewhere in-between as a blend between agile and predictive.

There are four development approaches (also called life cycles) on this continuum.

CHARACTERISTICS OF FOUR CATRGORIES OF LIFE CYCLES				
APPROACH	**REQUIREMENTS**	**ACTIVITIES**	**DELIVERY**	**GOAL**
Predictive	Fixed	One time	Single delivery	Manage cost
Iterative	Dynamic	Repeat till correct	Single delivery	Solution correctness
Incremental	Dynamic	One time per iteration	Frequent smaller deliveries	Speed
Agile	Dynamic	Repeat till correct	Frequent small deliveries	Customer value

Figure 3: Characteristics of Lifecycles (Agile Practice Guide Table 3-1 Page 18)

Predictive Life Cycles

Predictive life cycles are generally used on projects where the product to be produced is well understood. Scope, time and cost are defined early in the project. Predictive life cycles are also known as fully plan driven life cycles in which the project proceeds through sequential or overlapping steps. Each phase focuses on accomplishing a subset of project management processes and activities.

Iterative Life Cycles

In these life cycles, project phases intentionally repeat project activities as project understanding increases. Each successive increment adds to the deliverable's functionality and the product is developed in iterations which may or may not overlap. This approach allows feedback to improve and modify unfinished work.

Incremental Life Cycles

Project work is executed in a series of iterations or sprints and produces finished deliverables for immediate use.

Adaptive Life cycles

Adaptive life cycles are also known as change-driven or agile methods. This life cycle is used to respond to higher levels of change on a project and it involves ongoing stakeholder involvement. Several processes are typically performed in each iteration of this life cycle. Agile life cycles are both iterative and incremental and enable the frequent refinement and delivery of work items.

The overall scope is decomposed into a set of requirements (product backlog items) which will be worked on. At beginning of each iteration, the team determines which items can be delivered. The product is made ready for review by the customer at end of each iteration.
The following table is a summary of the four life cycles. Be sure to understand the characteristics of each lifecycle.

Hybrid Life Cycles

This is a combination of both predictive and agile life cycles. Parts of the project with well-defined requirements are managed using a predictive approach while those with evolving requirements are managed with an adaptive life cycle.

In actuality, predictive, incremental, iterative or agile approaches should not be viewed as "good or bad". Both could be used together or in isolation depending on the unique situation. When we use a bit of both, we call that a hybridized approach. The specific way we hybridize approaches is unique to every project.

A project team could decide to use a hybrid approach that is part agile and part predictive in different proportions. This could be done sequentially for some or all parts of the project as shown in the next four images.

FIGURE 4: AGILE DEVELOPMENT FOLLOWED BY A PREDICTIVE ROLLOUT (APG PAGE 27)

Hybridization could also be done concurrently for parts of the project as shown in the next image.

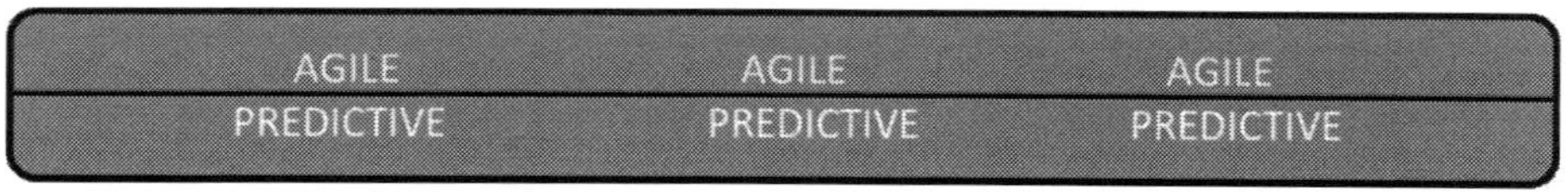

FIGURE 5: A COMBINED AGILE AND PREDICTIVE APPROACH (APG PAGE 27)

Alternatively, teams could inject a little bit of predictive into an agile project as shown in the next image.

Figure 6: A Largely Agile Approach with a Predictive Component (APG Page 28)

Alternatively, you could inject a little bit of agile into a largely predictive project resulting in a custom hybrid model as shown in the next image.

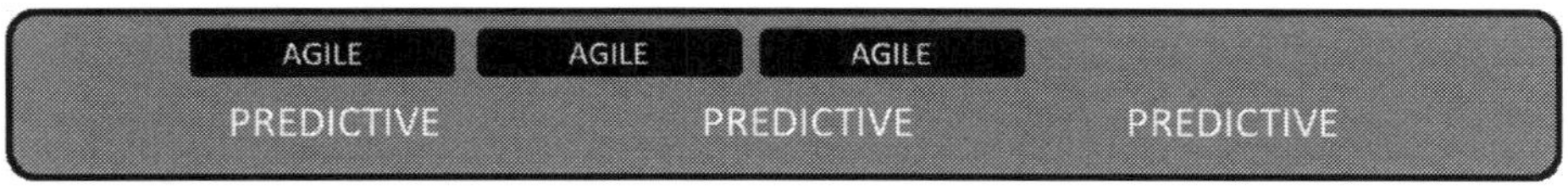

Figure 7: A Largely Predictive Approach with Agile Components (APG Page 28)

As project management is evolving, project teams and should have a wider, more inclusive view of the world of possibilities to be hybridized.

AGILE IS NOT JUST FOR IT

Let's talk about how Agile is not just for IT. That is where it started but consider predictive approaches started in construction, and we're using predictive approaches for pretty much any kind of project now. The same thing is going for Agile, which started in IT because it fits well there. Technology was changing very fast with the onset of object-oriented programming. It made it a lot easier to change things very quickly, and Agile fits well in that space. Well, since then Agile has grown substantially, it became mainstream, as you I'm sure you recognize, and we're using it everywhere. And now, there are adjustments, of course. What we're going to talk about is frequent change and the ability to redo things. And when you're dealing with physical products, that's not quite as easy, but it doesn't mean that Agile is not possible in the world where you're doing physical products. I do a lot of work in that space. I do a lot of work in the pharmaceutical industries, and also in organizations, we're building physical products, like firefighter's safety equipment, etc. We both have some experience in that space with Agile.

I want you to remember that IT or Agile is more about the mindset of how we do work rather than building in all of that flexibility and the way we do things in software. Some things are just not possible in all markets, consider that. And let's talk about some core concepts, some core practices that you can make use of regardless of the process. What you're going to learn as we get a little bit deeper into this is that there are a lot of good things that we've taken from the traditional approach and there are a lot of good things that we've implemented in Agile that can be used inside of a predictive (traditional) projects as well. Doing things like daily scrums, retrospectives, some of these kinds of short releases, all of these things are valuable regardless of the process that you follow.

Try to view this in the following sessions as opportunities to improve the way that you work today. We're going to be talking about several practices that you could make use off right away immediately even if you don't change a lot of your current process. So keep that in mind as we go through.

One thing that is stressed significantly in the Agile in a lean environment is the concept of continuous improvement or process improvement. It's not just about our process as it could be a significant aspect of it. It's also about ourselves, our team, our relationships with each other. How do we get better as an organization? And that's something that should transcend the practices that we're following. You don't have to be Agile to be looking at how to improve as an organization. That's something that we should be doing regardless of what practices were following. So these are some core concepts that we can make use of again in any method you follow.

Another thing that we're looking for is a way to eliminate the waste in your organization. You can think of at least one thing that you do as an organization that is not bringing you a lot of value. Every organization has waste and what we want to do is to find ways of getting rid of that waste or at least reducing it to the point where it's no longer causing significant challenges. We should always be asking ourselves, is this bringing us value? And if it's not, maybe we should find alternatives or get rid of it altogether. That's another core concept but again, regardless of what practices you're following, whether you're doing traditional or Agile, that's something that we should take internally and start following as well.

QUESTION 1

Agile is best used when __________

A. There is low variability is needed on the project

B. Change is unlikely and static requirements are needed

C. Experimentation and discovery are needed for the solution

D. Little chance of change and high certainty

You will never hear us say that predictive is a flawed process and should never be used. There are tons of great practices that we've essentially stolen and used in the Agile environment. We don't throw things out just because it's predictive. We take what's reasonable, and maybe we use it in a slightly different way, but those practices are valuable, like managing risk; that's not something that stops just because you're doing Agile, but we want to be even better at managing risk. So absolutely. What we're saying is to choose what makes sense for the situation. Sometimes predictive (traditional) will make sense, and other times maybe you've got too much complexity or too much lack of knowledge or a lack of experience in a particular area, and in those cases, maybe Agile will work better for you.

Think about it this way; uncertainty and complexity is project risk. It invites project failure. If you don't know what it is you're building, or if you don't know how to build, and we know there are all kinds of uncertainty, so Agile is built to help us resolve uncertainty. If you think about a predictive project and uncertainty, what we typically do is a prototype, or we do some research and analysis to get better information to learn why Agile is built around that process of constant experimentation, that allows us to learn. It will enable us to reduce the risk frequently, so in those environments where there's a lot of uncertainty or a lot of complexity, a lot of risk Agile works exceptionally well.

ANSWER: ***Experimentation and discovery are needed for the solution***

PROJECT SUCCESS SLIDERS

Project Success Sliders allow you to gauge priorities regardless of the process you're using. Their purpose is to help you identify what truly matters to you, whether it's related to a specific project or the organization as a whole. Here's how it functions: consider what holds importance for you. Each slider on the right-hand side indicates the level of importance. A rating of 1 signifies minimal importance, while a 5 denotes utmost significance. Focus primarily on what's most crucial.

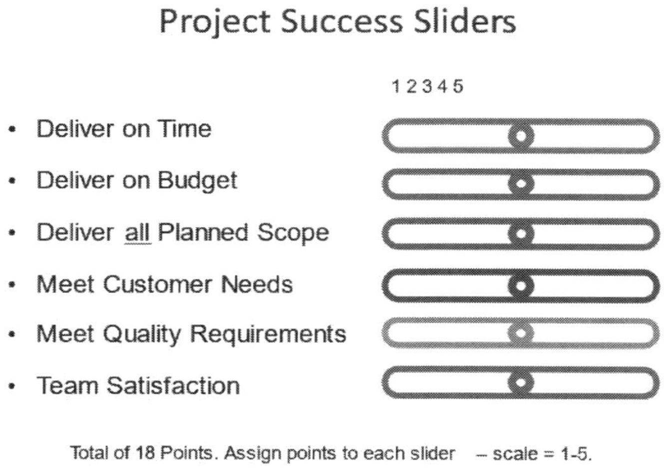

Figure 8: Project Success Sliders

The Project Success Sliders are designed to help you prioritize aspects of a project. Here's how they work:

1. **Allocation of Points**: You have a total of 18 points to distribute among six sliders. Each slider starts at a default value of 3, totaling 18 points (3 x 6 = 18). If you increase the value of one slider, you must decrease another to ensure the total remains 18.
2. **Purpose**: This exercise aims to help you reflect on what's crucial for a project or from an organizational standpoint. The six categories are:

- Deliver on time
- Deliver on budget
- Complete all planned scope
- Meet customer needs
- Fulfill quality requirements
- Ensure team satisfaction

Take a moment to consider which of these are most important to you personally and potentially to your organization.

3. **Observations**: In many instances, especially among project management professionals, the top three constraints (time, budget, and scope) are deemed vital. However, delivering within these constraints doesn't always equate to meeting customer needs. There are instances where projects are completed on time and within budget, but the customer remains unsatisfied. This discrepancy can arise if the project doesn't truly address the customer's needs or if there was a misunderstanding of the problem.
4. **Customer Needs**: In Agile, it's encouraged to prioritize customer needs. The term "customer" can be broad, but it generally refers to the individual or entity using or purchasing the product. It's paramount to understand and meet their needs, as delivering what's asked doesn't guarantee it's the right solution.
5. **Quality Requirements**: Quality is intrinsically linked to customer satisfaction. Delivering a subpar product can lead to significant dissatisfaction or even catastrophic outcomes, especially in life-critical sectors. Quality should never be compromised, even if time or budget constraints are pressing. In the IT realm, quality often gets sidelined, which can be detrimental.
6. **The Importance of Team Satisfaction**: A contented and motivated team is more likely to produce high-quality results, deliver projects on time, and adhere to budget constraints. Despite its significance, team satisfaction often ranks low on the priority list. However, it's essential to recognize that team morale directly influences delivery, quality, and other project outcomes. A satisfied team can significantly enhance the overall success of a project.

7. **Re-prioritizing**: In an ideal scenario, customer needs, quality requirements, and team satisfaction should be at the top of the priority list. While time, budget, and scope remain crucial, it might be beneficial to reassess their relative importance. In an Agile environment, there are strategies to ensure that projects are still delivered on time and within budget while emphasizing these priorities.
8. **Rethinking Scope**: The term "all planned scope" is emphasized for a reason. It's common in traditional project management to have a set of requirements that the project must fulfill. However, it's worth questioning the necessity of each requirement. Are all of them crucial for the project's success? Could some be postponed to a later iteration or release? It's essential to adopt a flexible approach, allowing for adjustments based on the project's evolving needs.

In summary, while traditional metrics like time, budget, and scope are undeniably important, it's crucial to ensure that team satisfaction, customer needs, and quality are not sidelined. Adopting a more holistic approach can lead to more successful and efficient project outcomes.

THE AGILE MANIFESTO – VALUES

Let's journey into one of the cornerstones of the Agile community: The Agile Manifesto.

1. **Nature of the Agile Manifesto**: The Agile Manifesto isn't a step-by-step guide on how to implement Agile. Instead, it presents a set of values and principles that guide our actions and interactions. It shapes our behaviors and how we collaborate.
2. **Comparison with PMBOK®**: Many might be familiar with the PMBOK® (Project Management Body of Knowledge), a comprehensive guide that aims to cover every conceivable project management scenario. While the PMBOK® is extensive, Agile doesn't seek such an exhaustive resource. The reason being, Agile values flexibility. It's expected that Agile practices will vary depending on the organization, project, or even the team. This adaptability might seem daunting to some, but it's essential to understand that even methodologies like the waterfall, which aim for standardization, are often implemented differently across organizations.
3. **The Role of the Manifesto**: Since Agile doesn't provide a detailed playbook, the manifesto serves as a compass. When faced with decisions or potential changes to the process, teams can refer to the manifesto to gauge if they're aligned with Agile values. It's a guiding tool, shaping the culture and behaviors within an organization. The manifesto begins with the acknowledgment of continuously

discovering better ways of product development by practicing and assisting others. It emphasizes valuing certain aspects over others, but it's crucial to note that it doesn't dismiss the latter.

4. **Misinterpretations**: The manifesto's phrasing has led to various myths and misconceptions. While it states a preference for the elements on the left over those on the right, it doesn't imply that the items on the right are unimportant or irrelevant.

In essence, the Agile Manifesto serves as a beacon, guiding teams and organizations in their Agile journey, ensuring they remain true to its core values and principles.

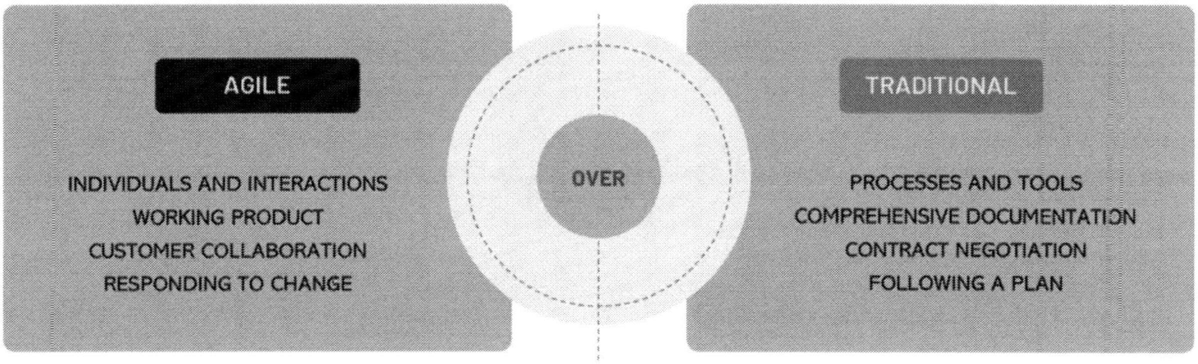

That is, while there is value in the items on the right, we value the items on the left more.

Let's examine the core values of the Agile Manifesto:

1. **Individuals and Interactions over Processes and Tools**: While Agile acknowledges the importance of processes and tools, it places a higher value on the individuals involved and their interactions. Consider this: given a choice between an exceptional team with average tools and an average team with exceptional tools, the preference would lean towards the former. A stellar team can navigate around mediocre tools and processes, achieving goals despite challenges. The essence here is that tools and processes should serve the team, not vice versa.

2. **Working Product over Comprehensive Documentation**: This value has sparked numerous misconceptions. Agile doesn't dismiss documentation. Especially in regulated industries, documentation is crucial. However, Agile emphasizes producing the right documentation at the right time. Instead of traditional documentation, Agile might employ dynamic tools like backlogs. The primary focus is on delivering a valuable product, with documentation supporting this goal, not overshadowing it.
3. **Customer Collaboration over Contract Negotiation**: While contracts or requirement lists are essential, Agile prioritizes understanding the underlying needs of the customer. It's about diving deep into the problems customers aim to solve rather than just ticking off items on a requirement list. Engaging in meaningful dialogues with customers to grasp the root cause of their needs is more valuable than merely adhering to a contract.

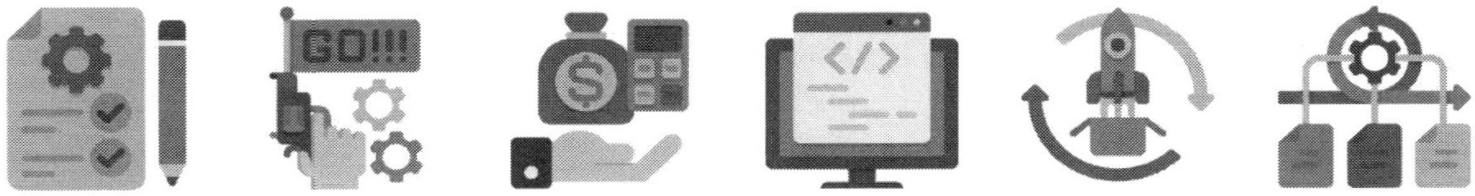

Often, when customers articulate their needs, they might not have a clear or accurate representation of what they truly desire. They might present a perceived solution, but it might not necessarily address their core needs. Instead of merely accepting their initial requests, it's essential to engage in a deeper dialogue. By respectfully asking questions like "Why do you need that?" or "What benefits are you seeking?", we can uncover their underlying needs and challenges. This approach, which aims to understand the root cause rather than just the surface-level request, is a proactive risk reduction technique. It ensures that we address the actual problem rather than just implementing a potentially misaligned solution

4. **Responding to Changes over Following a Plan**: A common misconception about Agile is that it's chaotic and lacks planning. This is far from the truth. Agile involves continuous, adaptive planning. While there's initial planning, more detailed planning unfolds as the project progresses, adapting to new insights and challenges.

The Cone of Uncertainty: Those familiar with project management, especially PMPs, might recognize the term "cone of uncertainty." It represents the idea that at the beginning of a project (visualized as the wide mouth of a funnel), there's the most uncertainty and variability. As the project progresses, this uncertainty narrows. In Agile, initial planning happens at the cone's mouth, where there's the least clarity. As the project advances and more is learned, planning becomes more detailed, ensuring the project remains on the right track.

Anticipating Changes: Agile is designed to accommodate changes. Instead of viewing changes as disruptions, Agile sees them as inevitable parts of the project landscape. The methodology is built to handle changes in a streamlined manner, without causing upheaval. Instead of rigidly adhering to an initial plan, Agile recognizes that plans might need adjustments as new information or challenges emerge.

"Everyone has a plan until they're punched in the face!"

- **Mike Tyson**

Adapting to Reality: As the Mike Tyson quote suggests, even the best-laid plans can face unexpected challenges. In the face of unforeseen obstacles, it's more pragmatic to adapt our plans rather than stubbornly sticking to them.

In essence, Agile is about navigating the unpredictable waters of project development with flexibility and foresight, always ready to adapt when reality throws a curveball.

QUESTION 2

The Agile Manifesto states:

Individuals and interactions over _____

A. Processes and tools
B. Comprehensive negotiation
C. Responding to change
D. Following a plan

The essence of successful project management and product development lies in the people involved and their interactions. While processes and tools play a role, they are secondary to the human element. Consider this: meeting deadlines, staying within budget, and adhering to the planned scope are outcomes that naturally arise from having a cohesive team that understands customer needs and is committed to delivering quality. Historically, many organizations prioritized shareholder value above all else, sometimes leading to questionable decisions. However, a shift in perspective emphasizes the importance of the team and customer-centric approaches, recognizing that genuine value and sustainable success stem from prioritizing people and their collaborative efforts.

Answer: ***Processes and tools***

QUESTION 3

The Agile Manifesto states:

A. Processes and tools
B. Comprehensive documentation
C. Responding to change
D. Following a plan

The importance of documentation in product development cannot be understated. It serves as a crucial reference, ensuring clarity and consistency. However, it's essential to strike a balance. While documentation is vital, it shouldn't overshadow the primary goal: delivering a valuable product to the customer.

Consider this analogy: No customer would prioritize documentation over the actual product. They won't be content with just a manual if they don't receive the product it describes. The product drives profit and satisfies customer needs, while documentation supports its creation and usage.

Many organizations, perhaps from past experiences or industry norms, lean heavily on documentation. But it's a misconception to equate the completion of documentation with progress on the product. A finished document doesn't necessarily translate to a product milestone. The true measure of progress is the product's development stages, not the accompanying paperwork.

Furthermore, while customers provide requirements, they might not always capture their true needs accurately. Open dialogue and collaboration with customers are crucial. It ensures that the end product aligns with their actual needs, rather than just ticking off items on a requirement list. The ultimate goal is customer satisfaction, and understanding their genuine needs is the pathway to achieving it.

Answer: Comprehensive documentation

QUESTION 4

Which of the following is **most** important on all projects?

- Deliver on Time
- Deliver on Budget
- Deliver all Planned Scope
- Meet Customer Needs
- Meet Quality Requirements
- Team Satisfaction

A. Time

B. Budget

C. The 3 lower-level items on the list

D. The 3 higher level items on the list

Engaging in open discussions, especially on contentious topics, is crucial. It fosters understanding and clarity. The real essence of delivering on time, within budget, and according to the planned scope is rooted in having a cohesive team that understands customer needs and is committed to producing a high-quality product.

Historically, many organizations prioritized shareholder value, sometimes at the expense of other critical aspects. In their quest to maximize shareholder value, they might rush products, compromise on quality, or even lay off staff. Jack Welch of GE, once a proponent of this approach, later admitted its flaws. He realized that a company's primary objective isn't merely profit generation. Instead, it should focus on acquiring and retaining customers. When a company prioritizes customer satisfaction, profit naturally follows.

This perspective shift emphasizes outcomes over outputs. While delivering on time and other such metrics are outputs, understanding and meeting customer requirements are outcomes. The latter should be the primary focus.

However, it's often challenging to align customers, especially in vendor-client relationships, with Agile principles. Traditional contracts, with their fixed scopes and prices, can conflict with Agile's flexible nature. Ideally, customers would fully grasp and support Agile methodologies, but this isn't always feasible. Compromises might be necessary, and contracts might need adjustments to align better with Agile while still providing clients a sense of control. The ultimate goal is to strike a balance that benefits both parties.

Answer: The 3 lower-level items on the list

The Agile Manifesto

The Agile Manifesto, while succinct in its values, is further elaborated upon through its set of principles. These principles provide a deeper understanding and context to the values, offering a more comprehensive guide for organizations to follow.

However, it's essential to note that the manifesto and its principles aren't prescriptive blueprints for Agile practices. Instead, they serve as a philosophical foundation, emphasizing the desired culture and behaviors within an organization. The principles hint at practices but don't dictate specific methods or techniques for product development or project management. Their primary focus is on the "why" behind the actions, encouraging organizations to internalize the underlying reasons and motivations rather than merely following a set of rigid steps.

In essence, the manifesto and its principles aim to instill a mindset and culture that prioritize collaboration, adaptability, and customer satisfaction, providing a framework for organizations to navigate the complexities of product development and project management.

The Agile Manifesto
12 Principles

01 Our highest priority is to satisfy the customer through early and continuous delivery of valuable product.

02 Welcome changing requirements, even late in development. Agile processes harness change for the customer's competitive advantage.

03 Deliver working product frequently, from a couple of weeks to a couple of months, with a preference to the shorter timescale.

04 Business people and developers must work together daily throughout the project.

05 Build projects around motivated individuals. Give them the environment and support they need, and trust them to get the job done.

06 The most efficient and effective method of conveying information to and within a development team is face-to-face conversation.

07 Working product is the primary measure of progress.

08 Agile processes promote sustainable development. The sponsors, developers, and users should be able to maintain a constant pace indefinitely.

09 Continuous attention to technical excellence and good design enhances agility.

10 Simplicity – the art of maximizing the amount of work not done – is essential.

11 The best architectures, requirements, and designs emerge from self-organizing teams.

12 At regular intervals, the team reflects on how to become more effective, then tunes and adjusts its behavior accordingly.

Now, you can read through them yourselves as well and kind of interpret some of them, but there are a few of them that I really want to point out.

1. Our highest priority is to satisfy the customer through early and continuous delivery of valuable products.

This principle stands out prominently among the rest. While all principles in the Agile Manifesto are crucial, the authors emphasized this one by placing it at the forefront. The essence of this principle isn't just about delivering products early and continuously; it's about the overarching goal of satisfying the customer.

Every action, every decision, and every piece of work should be viewed through the lens of customer satisfaction. It's a reminder that, at the end of the day, the success of any endeavor is gauged by how well it meets the needs and expectations of the customer.

However, this doesn't mean blindly following every customer request. It's about understanding their true needs, even if it means guiding them away from what they initially thought they wanted. The aim is to ensure that the end result aligns with their genuine requirements and brings them value. It's not merely about the adage "the customer is always right," but rather about doing what's genuinely best for the customer to ensure their satisfaction.

2. *Welcome changing requirements, even late in development. Agile processes harness change for the customer's competitive advantage.*

This principle can be a bit controversial, and that's precisely why it's intriguing. The idea of allowing changes even late in product development can be daunting for many. However, if we embrace change throughout the project in a streamlined manner, the likelihood of significant changes towards the end diminishes. Implementing a cumbersome change management process can lead to resistance, causing delays and larger, more impactful changes.

It's not to say we shouldn't manage change. Especially in the IT realm, it's more feasible, but with physical products, it's a bit more challenging. The key is to foster an environment where change is expected and managed efficiently. This way, changes can be integrated throughout the project's lifecycle, preventing massive, costly alterations at the end.

The last thing we want is to present a product or results to our customers and then face significant changes. By allowing for continuous adjustments, the overall project becomes more manageable and less prone to substantial late-stage modifications.

3. *Deliver working software frequently, from a couple of weeks to a couple of months with a preference to the shorter timescale.*

I want to emphasize that the feasibility of rapid delivery varies based on the nature of the product. While some products, especially in software, can be delivered swiftly, others, especially physical products, require more time. The essence of this principle is to encourage delivery in small increments as quickly as feasible, primarily to gather feedback. The term "deliver" can be flexible. In the context of physical products, it might mean a demonstration or prototype, while in IT, it could mean actual product deployment.

Take Amazon as an illustration. They release changes into production over a hundred times daily. This frequency is possible due to their high level of automation in both verification and deployment processes. However, it's essential to understand that not every organization can or should emulate Amazon, especially those in the physical product domain.

The overarching goal is to present or "deliver" work regularly to stakeholders or customers. This frequent showcasing serves as a feedback loop, allowing teams to adjust based on real-world reactions. Think of it as a series of experiments: you present a version of the product, gather feedback, and refine. This approach ensures that the final product aligns more closely with user needs and expectations, rather than just meeting initial specifications.

4. *Business people and developers must work together daily throughout the project.*

Why is this essential? Continuous collaboration is key to understanding and adapting to changes efficiently. It allows the team to self-correct and stay aligned with the project's goals. In Scrum, the most widely adopted Agile framework, the Product Owner acts as the representative for the business or customer. Their continuous interaction with developers facilitates a rapid feedback loop, enabling swift adaptations. The Product Owner provides clarity on customer requests and requirements, becoming an invaluable resource for developers as they shape the product.

5. *Build projects around motivated individuals. Give them the environment and support they need and trust them to get the job done.*

This principle can be applied universally across any project, whether it's rooted in Predictive or Agile methodologies. Arguably, it's the most pivotal organizational principle in the entire set. It underscores the importance of starting with exceptional individuals—those who are passionate about their roles, take pride in their work, and can operate efficiently with minimal to no oversight. The challenge, however, isn't just in hiring these motivated individuals but in maintaining their motivation. Even if an organization manages to recruit the best talents, without a conducive environment that sustains their motivation, their enthusiasm can wane, leading to increased oversight to ensure job completion.

It's crucial to emphasize the hiring of mature, responsible adults and then providing an environment where they remain motivated. This dual approach—hiring dedicated individuals and fostering a supportive environment—creates a foundation of trust. When teams are trusted, there's less need for constant oversight and direction, allowing for greater creativity and innovation. This cultivates a culture of engagement, essential for knowledge work.

While I may not be privy to the intricacies of your organization's culture, this principle is universally applicable, especially in Agile practices. In Agile, there's an inherent trust in teams, allowing them to operate without constant interference. This approach aligns with the concept of servant leadership, where leaders serve their teams. The role of leadership evolves from merely directing tasks to setting a clear vision and mission, explaining the objectives and reasons behind them, and then asking, "What do you need from me to achieve this?" This shift in leadership style, although challenging, is integral to fostering a collaborative and efficient work environment.

6. The most efficient and effective method of conveying information to and within a development team is face-to-face conversation.

The principle emphasizes the unparalleled value of face-to-face communication. Historically, this method was often taken for granted. However, recent shifts to remote work and online collaboration have highlighted the challenges of communicating without the benefit of physical presence. This has led to a surge in webcam purchases, as people strive to maintain some semblance of face-to-face interaction. The spike in demand for webcams underscores a collective realization of their importance in facilitating effective communication.

The importance of face-to-face communication is underscored by Professor Emeritus Albert Mehrabian's research. In situations where words, tone of voice, and body language are incongruent, especially in high-stakes communications, the receiver often relies on non-verbal cues to decode the true intent. Mehrabian found that a mere 7% of the empathetic communication is attributed to the actual words spoken. In contrast, body language accounts for a significant 55%.

This research highlights the profound impact of visual cues in communication. It provides context to the recent surge in webcam purchases and the booming popularity of platforms like Zoom. While virtual meetings cannot fully capture the depth and nuance of in-person interactions, they come close by allowing participants to see each other. This aligns with the sixth Agile principle, emphasizing the value of face-to-face (or camera-to-camera) communication in our increasingly digital world.

7. ***Working product is the primary measure of progress.***

Now, this kind of goes back to the concept of less documentation and more focus on product. In Agile, the primary indicator of progress is not the completion of documentation or reaching certain milestones in a phased approach, but rather the delivery of working products. This principle emphasizes the tangible results over theoretical progress.

Traditional project management methods, like predictive approaches, often measure progress in terms of phases completed. For instance, finishing the requirements phase or the design phase might be seen as achieving a certain percentage of project completion. Agile challenges this notion. Instead of relying on such phase-based metrics, Agile focuses on the actual product being built.

This shift in perspective might seem subtle, but it's profound. In Agile, there's a binary view of progress: a feature or product is either done (meaning it's fully functional and meets its acceptance criteria) or it's not. There's no in-between, no "almost done" or "partially done." This clarity eliminates ambiguity and ensures everyone has a shared understanding of what "done" truly means.

While Agile does allow for tracking and metrics, the emphasis is different. Instead of measuring how much of a document is completed, the focus is on how much of the product is functional and meets the user's needs. This approach provides a more accurate and meaningful gauge of progress. As we venture deeper into Agile, we'll explore the various metrics used and how they offer a clearer picture of a project's status compared to traditional methods.

8. Agile processes promote sustainable development. The sponsors, developers, and users should be able to maintain a constant pace indefinitely.

In essence, sustainable work practices prioritize quality and team well-being over sheer output. When teams are pushed to their limits with overtime, not only does the quality of work deteriorate, but team morale and satisfaction also take a hit. This can lead to burnout, increased turnover, and a decline in overall productivity in the long run.

The concept of "velocity" in Agile is about understanding a team's consistent rate of delivering value. It's not about pushing the team to constantly increase that rate, but rather to find a steady, sustainable pace. This ensures that the team can maintain their quality standards and stay engaged and motivated over the long term.

Moreover, when teams are overworked, they tend to make more mistakes, leading to more rework and delays. This is counterproductive and goes against the Agile principle of delivering valuable products efficiently.

In a sustainable Agile environment, the focus shifts from merely meeting deadlines to ensuring that the work being done is of the highest quality and that the team remains cohesive, motivated, and satisfied. This approach not only benefits the team but also results in better products and happier customers.

QUESTION 5

In the Agile Manifesto, our highest priority is to ______

A. Satisfy the CEO

B. Satisfy the customer

C. Satisfy the project team

D. Make as much profits

A lot of companies I work with, they really don't focus on the customer. They're more focused on delivering on time on budget because that defines success. I'll never understand that concept. They're important but that's not success.

***Answer:* Satisfy the customer**

QUESTION 6

In the Agile Manifesto, we should deliver working product frequently, from __________, with a preference to the shorter timescale.

A. A couple of weeks to a couple of months

B. A couple of days to a couple of weeks

C. A couple of months to a couple of years

D. A couple of years to a couple of decades

This is where the trick question comes in, so the manifesto specifically states a couple of weeks to a couple of months, but the reality is if you can deliver things in a couple of days or a couple of weeks awesome! Really, the goal here is not the timebox. It's more about how frequently you can deliver high-quality product to get that feedback and maybe in some cases get some incremental return. This is a memorization question. If you're going to need to answer this on a quiz, A is the correct answer, but B is somewhat correct as well in terms of real-world application (but again, we cannot restrict delivery to a couple of weeks only). I wish to say that none of you are really wrong, if you're internalizing it but the correct answer is A.

Answer: A couple of weeks to a couple of months

QUESTION 7

In the Agile Manifesto, business people and developers must work together __________

A. At key points in the project

B. Daily throughout the project

C. Weekly in the project

D. Monthly in the project

This is a good one. Well done because this is not one that we specifically called out as we were discussing. So the correct answer is daily throughout the project.

it's a it's not

Answer: Daily throughout the project

QUESTION 8

In the Agile Manifesto, we build projects around______

A. Motivated individuals

B. Babysitting teams

C. Theory X individuals

D. Carrot and stick approach

A Theory X management approach in essence believes that people cannot work without supervision and micromanagement. Theory X is that, essentially, that the organization doesn't think that people want to do their jobs. It's the view that people need to be controlled and need to be managed and need to be told what to do, then theory Y is that people generally want to do their jobs and it should be leadership's job to make sure they have the ability to do so. It's a very different approach. Most organizations are Theory X. That's why they have managers in teams. That's where they tell people and have all these control points in place, most organizations are Theory X.

The concept there is how do we motivate people. Now, in the past, it was viewed as the carrots and sticks to approach, right? If somebody does something cool, give them a bonus, right? Give them a couple grand or something like that, and that makes them happy. Well, money does motivate of course, but what motivates more are some deeper internal things. Things like being able to direct yourself, making your own decisions, learning new things. It's that motivation factor that is beyond dollars, beyond carrots and sticks, and that's what we really want in a creative, innovative, knowledge-based kind of environment. That's the approach, that's what we really want for motivation.

Answer*: *Motivated individuals

QUESTION 9

In the Agile Manifesto, the most efficient and effective method of conveying information to and within a development team is ______

A. Online communication

B. Information radiation

C. Face-to-face conversation

D. Email

Obviously, today remote teams are kind of the norm and that might be the case for quite a long time.

We absolutely prefer, and I think we can all recognize, that face-to-face conversation is obviously the most effective. There's no question about that. You get the body language, you get the facial expressions, and all those things. That's how we interact and that's how we gain a huge amount of information as we're talking with each other. Unfortunately, it's not as possible, so the online communication is definitely part of our reality, but at least if we can do the webcam so we get more of the facial expressions, that definitely helps us a lot. The correct answer from the manifesto perspective is face-to-face conversation.

The example like to give on this one is texting. We all text. I don't know about you but when I'm texting my certain family members, I am getting myself in trouble a lot, because I'll give a short answer because I'm busy and then they ask "how are you doing?" and I'll respond saying "fine" and they'll think I'm angry and it's because I can't express emotion using just words because there's no tone of voice. I'm not blaming them in any way, this is just communication right? So we want to have the richest form of communication so that we can truly understand not just the words that are being said but the emotion behind them.

Absolutely, we strongly encourage the video, the webcams, we all have them now, right? Again, this is the new norm. We all got webcams. We may not like them. Sometimes you might want to be working in your pajamas or you know, fuzzy bunny slippers or something like that, but you know, face-to-face is webcam now and that's what we got to have to get used to.

Answer: Face-to-face conversation

QUESTION 10

Agile processes promote ___________

A. Sustainable development

B. Sustainable management

C. Sustainable forests

D. Sustainable energy

At least according to the manifesto, the correct answer would be A, sustainable development. And the concept behind this is that we don't push people to do more than they're capable of doing and we're going to track that, we're going to look at the historical data that says on average, this team is able to do this amount of work every iteration or every timebox. We'll talk more about that soon. So we measure that so that we're not constantly pushing the team harder than they had the capacity to do the work. Again, the intent behind this is that if you push the team to do more than they're capable of doing, it causes all kinds of bad behaviors. It causes poor quality, it causes people being tired. I know a guy who had a heart attack because he was working too many hours. Healthcare challenges can happen with this. So what we want to do is plan for reality, not for desire and hope.

It's kind of funny, I was working with one team and I was talking about this and one person said, "We're going to call you the hope killer." I wasn't sure how to take that, but you know, then I realized what he was really saying was that we don't plan for hope. We hope we can get something done. That's not a plan, right? That's a wish, all right? So give that consideration, want to make sure you're planning for a reality.

Answer: Sustainable development

9. Continuous attention to technical excellence and good design enhances agility.

The summary is that following good engineering and quality management processes and practices will reduce wasted time and effort, hence making you more lean and more agile.

Quality Over Quantity: Prioritizing technical excellence means that the team is focused on producing high-quality work rather than just completing tasks. This reduces the likelihood of defects and rework.

Foundation for Future Work: A well-designed system is easier to modify and extend. As business needs evolve, a solid foundation ensures that the system can adapt without requiring a complete overhaul.

Efficiency: Good design and technical excellence mean that the system is optimized and efficient. This reduces resource wastage and ensures that the system can handle the demands placed on it.

Reduced Technical Debt: By prioritizing technical excellence, teams can avoid accumulating technical debt – the "cost" associated with postponing necessary upgrades or fixes. Over time, technical debt can slow down teams and make changes more challenging to implement.

Faster Time-to-Market: With a focus on quality and design, teams can streamline their processes, reducing the time it takes to bring a product or feature to market.

Improved Collaboration: A well-designed system with clear documentation and standards makes it easier for team members to collaborate. They spend less time trying to understand the system and more time improving it.

Customer Satisfaction: Ultimately, a high-quality product leads to higher customer satisfaction. Customers benefit from a product that works efficiently, reliably, and meets their needs.

10. Simplicity – the art of maximizing the amount of work not done is essential.

Avoid engaging in unnecessary tasks just because they're outlined in a contract. This is where Agile diverges from the Predictive world. In Predictive environments, if it's in the plan or part of the scope, it's deemed essential. Everything outlined has to be executed. This is why we use plan-driven approaches for specific scenarios, ensuring that what's in the scope gets done. However, in the Agile realm, where rapid adaptation is the norm, the perspective shifts.

For instance, in construction, you wouldn't decide to skip building the 6th storey and jump straight to the 7th, leaving out the 5th and 6th. It just doesn't work that way. But in Agile projects, such flexibility is possible and often encouraged. While there might be elements in a plan-driven project that can be adjusted or swapped out, it's not the standard approach. The emphasis in Agile is on delivering what's truly necessary and valuable, rather than strictly adhering to a predefined plan.

11. The best architectures, requirements, and designs emerge from self-organizing teams.

Allowing teams the autonomy to self-organize can lead to superior outcomes. Even in plan-driven projects, providing teams with some flexibility and discretion in specific tasks or phases can yield better results. Empowering teams in this manner often results in more innovative solutions and a stronger sense of ownership over the work

12. At regular intervals, the team reflects on how to become more effective, then tunes and adjusts its behavior accordingly.

Embracing the principle of continuous improvement is crucial. The term "Kaizen," which means continuous improvement, encapsulates this idea. While traditional project management approaches might incorporate "lessons learned" sessions, these are often held at the project's conclusion. While these sessions hold value, their benefits might not always be fully realized or implemented. In contrast, Agile promotes frequent reflection and adjustments throughout the project, ensuring that improvements are timely and directly impact the ongoing work. This proactive approach ensures that teams are consistently evolving and optimizing their processes.

Lessons Learned: The Traditional Approach

In many organizations, the approach to "lessons learned" can be somewhat flawed. Often, these sessions are conducted at the end of a project, when it's potentially too late to make impactful changes. The scenario usually unfolds like this:

1. **Timing**: The session is held after the project concludes, sometimes after a project has faced significant challenges or even failed.
2. **Cost**: These meetings can be costly, involving many team members and consuming significant amounts of time.
3. **Documentation**: Detailed notes are taken, capturing complaints, suggestions, and insights.
4. **Storage**: These notes are then compiled into a document, often stored in a location like SharePoint.
5. **Follow-up**: Unfortunately, in many cases, this document is rarely revisited. The insights captured don't influence future projects as they should.

The result? Valuable insights and potential improvements are lost. Each new project might repeat the same mistakes, and the cycle continues. While the intent behind "lessons learned" is commendable, the execution and follow-through often leave much to be desired.

Retrospectives: A Continuous Improvement Approach

In contrast to the traditional "lessons learned" sessions, many organizations are now adopting the practice of "retrospectives." This approach, while not exclusive to Agile, is a hallmark of Agile methodologies. Here's how it differs and why it can be more effective:

1. **Frequency**: Instead of waiting until the end of a project, retrospectives are conducted regularly, often every two weeks or at the end of each sprint.
2. **Focus**: The team discusses what went well and what didn't during the period in review.
3. **Actionable Insights**: Rather than just documenting issues, the team identifies one or two specific actions to improve in the next cycle.
4. **Continuous Improvement**: By addressing issues in real-time, the team can make immediate adjustments, leading to incremental improvements throughout the project's lifecycle.
5. **Time Investment**: Retrospectives are typically short, often lasting about an hour, making them a manageable commitment for teams.
6. **Team Dynamics**: For retrospectives to be effective, it's crucial for the team to foster an environment of trust, openness, and a genuine desire to improve.

In essence, retrospectives shift the focus from merely documenting issues to actively addressing them in real-time. This proactive approach ensures that teams are continually refining their processes, leading to more successful project outcomes. Even for teams not practicing Agile, incorporating regular retrospectives can be a game-changer in driving continuous improvement.

CHAPTER 2

AGILE FACTS

AGILE SELLING POINTS

When diving into the world of Agile, it's crucial to understand the distinction between merely "doing Agile" and truly "being Agile."

1. **Doing Agile**: This is the surface level of Agile. It's about adopting the practices and methodologies like Scrum or Kanban. It's the tangible actions and processes that can be easily observed and measured. However, merely adopting these practices doesn't necessarily mean you've embraced the Agile mindset.
2. **Being Agile**: This is the deeper, more profound level of Agile. It's about internalizing the principles and values of Agile. It's a shift in mindset, culture, and approach. Being Agile means you're not just following a set of practices; you're living and breathing the Agile philosophy. It's about fostering a culture of continuous improvement, collaboration, and customer-centricity.

While many organizations might adopt Agile practices because they see others doing it, that shouldn't be the primary motivation. The real value of Agile comes when it's embraced because it aligns with the organization's goals and values. It's not just about changing processes; it's about transforming the way you think and work.

In essence, while "doing Agile" might be your starting point, the ultimate goal should be "being Agile." This transformation might require changes in organizational culture and structure, but the benefits in terms of flexibility, responsiveness, and customer satisfaction can be immense.

WHAT'S IN IT FOR ME?

Now, the thing to remember is that Agile is mainstream. Most organizations today are following some form of Agile approaches today. Agile is beneficial for the following reasons:

1. **Agile is Mainstream**: Agile is no longer a niche methodology. It's widely adopted across industries and organizations. However, it's essential to note that while many claim to practice Agile, not all do it effectively.
2. **Debunking Myths**: There's a misconception that Agile is just the latest trend. In reality, Agile's roots trace back over three decades. The first Agile teams were established in the 1990s, and even before that, Agile-like practices were in place. Some of us recall implementing Agile-like strategies as far back as the 1980s.
3. **Agile's Longevity**: Agile isn't a fleeting trend. It's here to stay. Recognized institutions like PMI® have integrated Agile into their frameworks, exams, and guides.
4. **Personal Benefits of Agile**: Embracing Agile isn't just beneficial for organizations; it's advantageous for individuals too. In today's job market, having Agile expertise is almost a necessity. It's a sought-after competency that can significantly enhance your professional profile.
5. **Agile's Versatility**: Agile practices can enhance even traditional predictive projects. Techniques like Kanban, for instance, can be integrated into both Agile and predictive frameworks to optimize processes. Moreover, many Agile

principles and practices, such as retrospectives and daily stand-ups, can be universally applied to improve various projects.

6. **Agile ≠ Elimination of Project Management**: There's a misconception that Agile eradicates the need for project management. This isn't true. While Agile might shift some roles and responsibilities, the essence of project management remains. Complexity, risk, and other project facets still need managing.
7. **Approach with an Open Mind**: It's essential to approach Agile with an open mindset. Don't be swayed by every opinion on the internet. While Agile might change certain aspects of project management, it doesn't negate its importance. Embrace Agile for its strengths and understand its nuances to leverage it effectively.

Project management remains essential, regardless of the methodology in use. There's a misconception suggesting that Agile eliminates the need for project managers. This isn't accurate. In scenarios involving small, low-risk projects, the role of a project manager might seem less pronounced, but this holds true for both Agile and traditional predictive approaches. When projects scale up in complexity, especially those involving multiple teams, the expertise of a project manager becomes invaluable.

In an Agile environment, the role of the project manager evolves. It shifts back to what it was originally intended to be: managing projects, not people. Traditional methodologies often saw project managers directing individual tasks and overseeing team members closely. However, Agile promotes team autonomy, where team members manage their tasks. This allows project managers to refocus on their primary responsibilities, such as managing risks, handling complexities, fostering collaboration, and engaging stakeholders. In essence, Agile re-emphasizes the foundational principles of project management.

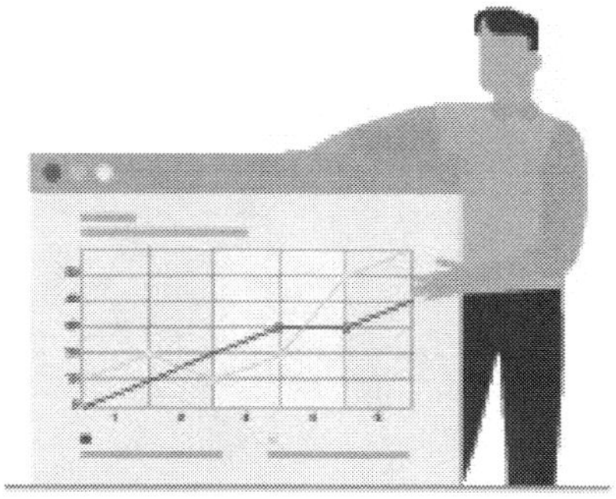

CONTINUOUS IMPROVEMENT

Continuous improvement is a cornerstone of Agile, encapsulated by the Japanese term "Kaizen," which translates to "good change" or "change for the better." This philosophy emphasizes the relentless pursuit of betterment, not just in processes but in all aspects of work and life.

While Agile formalizes this concept through retrospectives, inspired by the principle of reflecting regularly to enhance effectiveness, the spirit of continuous improvement should permeate every aspect of our work. It's not limited to these formal sessions. Daily stand-ups, interactions among team members, and even our personal endeavors should echo this ethos. While I strive for personal continuous improvement, it's not about perfection but the consistent effort to progress.

The team's evolution is as crucial as refining processes. It's about enhancing team dynamics, fostering better relationships, and refining practices. And who better to pinpoint areas for improvement than the team itself? There's a common misconception that managerial intervention is essential for a team's growth. However, empowering teams to make decisions fosters a deeper sense of ownership and engagement. When teams have autonomy, they not only take responsibility for their decisions but also feel more accountable for the outcomes. This sense of ownership is psychologically empowering and drives teams to excel. Thus, the focus should always be on evaluating and refining our processes, our team dynamics, and our individual contributions to foster an environment of relentless betterment.

IT'S ABOUT THE CUSTOMER

Agile places the customer at its core. Regardless of the methodology—be it Agile, predictive, or traditional —the customer's needs and satisfaction should always be the primary focus. Quality is paramount. While speed is essential, it should never come at the expense of quality. Delivering a high-quality product or service ensures customer satisfaction.

Many organizations prioritize output, using it as a success metric. However, focusing solely on output might not guarantee the desired outcome. Consider this: delivering one impactful feature that delights customers is far more valuable than producing ten features that don't resonate with them. While measuring output is essential to track progress, the emphasis should be on the outcome, even though it's more challenging to quantify. The goal is to ensure that the output aligns with the desired outcome.

Continuous improvement is a recurring theme in Agile. To truly satisfy customers, teams must be willing to adapt, evolve, and refine their approaches, relationships, and interactions.

Experimentation is a cornerstone of Agile. Teams are encouraged to try new approaches, assess their effectiveness, and then decide whether to continue (persevere) or change direction (pivot). This mindset requires organizations to embrace and encourage experimentation. It's not about labeling outcomes as "failures" or "successes." Instead, it's about learning and iterating. Every experiment, regardless of its outcome, provides valuable insights that inform future decisions. This iterative approach ensures that teams are always moving closer to delivering the right product or solution for their customers.

DEFINING SUCCESS

The traditional approach to project management revolves around the iron triangle or the triple constraints: scope, time, and cost.

Scope: In conventional project management, the scope is often predetermined. A requirements document is presented, outlining the expected deliverables. This scope is then considered fixed, and the project's success is often gauged by how well the team adheres to this predetermined scope.

However, this approach has its limitations. While delivering on the predetermined scope within the set time and budget is an achievement, it doesn't necessarily equate to project success. If the delivered product doesn't meet the customer's actual needs or if their needs change during the project's duration, then strictly adhering to the initial scope might result in a product that's not truly valuable to the customer.

This traditional approach to scope can be limiting. It assumes that all requirements are known and fixed at the project's outset. But in reality, needs can change, new information can emerge, and market conditions can shift. Rigidly sticking to an initial scope can lead to products that are out of touch with current needs.

In essence, while the iron triangle provides a framework for managing projects, it's essential to remember that the ultimate goal is to deliver value to the customer. Success isn't just about meeting predefined constraints but about ensuring that the end product aligns with the customer's actual needs and provides genuine value.

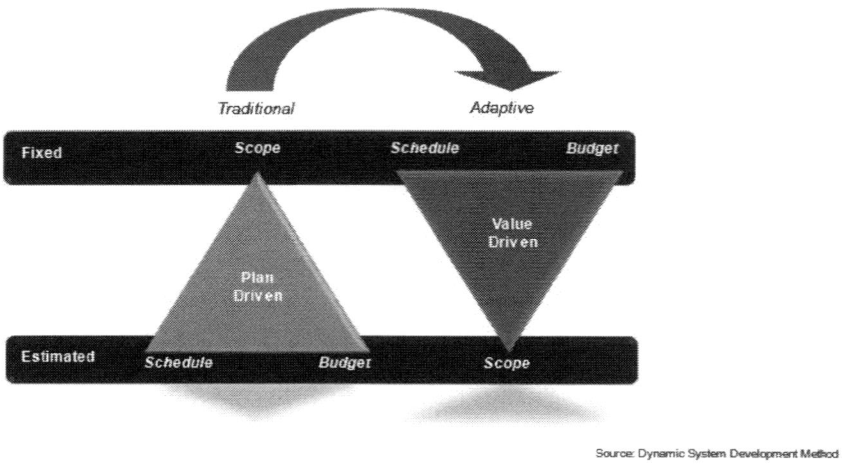

Figure 9: The Triple Constraint Reversed in Agile

In traditional project management, the iron triangle's three constraints—scope, schedule, and budget—are often set in stone. This rigidity can lead to challenges:

Many organizations start with a fixed set of requirements. Once these are established, the schedule and budget are estimated based on this scope. The problem arises when all three points—scope, schedule, and budget—become immovable. Ideally, one of these should remain flexible to accommodate changes, but this flexibility is often overlooked.

Quality: As the project progresses and it becomes evident that not all scope will be completed within the set schedule and budget, quality often becomes the sacrificial lamb. In the rush to meet deadlines, the quality of the deliverables may be compromised. This is especially problematic in industries where quality cannot be compromised, such as in the production of physical goods. In the software industry, this often results in subpar products being pushed to production, which can lead to customer dissatisfaction and additional costs down the line.

Schedule: The pressure to adhere to a fixed schedule often leads to overtime. While this might seem like a solution, it brings its own set of problems. Extended work hours can lead to burnout, decreased quality of work due to fatigue, and a demotivated team that moves to the next project already exhausted.

In the adaptive world of Agile, there's a shift in this approach. Instead of fixing the scope and adjusting the schedule and budget around it, Agile proposes the opposite. The scope becomes the variable, while the schedule and budget are fixed. This approach offers several advantages:

1. **Flexibility in Scope:** By treating scope as a variable, teams can prioritize the most valuable features and ensure they are completed to a high standard within the set schedule and budget.
2. **Timeboxes:** Agile operates within timeboxes, which are fixed periods during which specific tasks or features are completed. This approach naturally aligns with end-date scheduling, making Agile teams adept at meeting deadlines.
3. **Customer Value:** By focusing on delivering the most valuable features within the fixed schedule and budget, Agile ensures that the end product aligns more closely with customer needs and provides genuine value.

In essence, Agile's approach to the iron triangle prioritizes delivering value within a set timeframe and budget, allowing for flexibility in scope to ensure the highest quality output.

Budget considerations play a pivotal role in Agile methodologies. Here's a breakdown of the approach:

Budget: In Agile, there's a strong emphasis on having a fixed budget. This might seem counterintuitive initially, but it's grounded in logic. If you have a predetermined timeframe (schedule) and a consistent team size (no changes in team members), then your labor costs, a significant portion of the budget, become predictable. With fixed time and a stable team, the budget essentially becomes fixed.

Scope Estimation: With the schedule and budget set, the next step is to gauge the scope that can be achieved within these constraints. This might mean prioritizing essential features and possibly excluding others that can't be completed within the given time and budget. This approach emphasizes delivering the most value with n the constraints, rather than trying to cram in every possible feature.

Adjusting Scope: If it becomes clear that certain crucial features can't be completed within the set constraints, then there are decisions to be made. One option is to extend the schedule, which would also impact the budget. This decision should be based on the project's needs and the potential value of the features in question.

Increasing Capacity: Another option, typically considered as a last resort, is to increase the team size. However, this isn't a straightforward solution. Introducing new team members can increase complexity, introduce potential risks, and require time for the newcomers to get up to speed. This phenomenon is often referred to as the "mythical man-month" concept, which posits that adding manpower to a late project makes it later. Hence, increasing team size should be approached with caution.

Stable Teams: One of the foundational concepts in Agile is maintaining stable teams. When teams remain consistent, without frequent changes in members, it leads to better predictability in terms of schedule and budget. Stable teams develop a rhythm and understanding that can enhance efficiency and productivity.

In summary, Agile's approach to the iron triangle emphasizes setting clear constraints in terms of schedule and budget, and then working within those constraints to deliver the maximum value. This approach requires a shift in mindset, focusing on delivering value and quality over sheer quantity of features. It also underscores the importance of stable teams and the potential pitfalls of simply throwing more resources at a project to meet deadlines.

PUSH VS PULL

In work management, there are two main approaches: push and pull systems. Most organizations currently employ a push system, where work is assigned to individuals without proper prioritization. However, this approach often leads to inefficiencies, as the focus is on keeping people busy rather than working on the right tasks. It results in wasted time, money, and resources on unimportant work.

To address these challenges, we propose a pull system, which is more efficient and effective. Instead of assigning work to individuals, we create a list of available or prioritized tasks for the team. This promotes rigorous prioritization, ensuring that the most important work receives attention. Additionally, the pull system allows individuals to self-manage their capacity, eliminating the need for constant supervision. We trust the team to do their jobs and rely on transparency to verify their progress.

Transparency plays a crucial role in this process. By making everything the team is working on visible, we foster an open and honest environment. Empowering individuals to make decisions for themselves increases engagement, accountability, and ownership. Rather than pushing work onto people, we embrace a paradigm shift where the team volunteers for tasks.

It may take time for this concept to settle in, as it challenges the traditional approach. However, with a focus on visibility and transparency, we can demonstrate its effectiveness. We emphasize the importance of finishing tasks rather than starting many simultaneously. This is supported by the concept of Littles Law, which states that working on fewer tasks leads to faster completion. Therefore, the emphasis should be on finishing rather than starting. Pull systems help manage work in progress and will be a core concept in our sessions.

Overall, by adopting a pull system, organizations can streamline work management, prioritize effectively, and achieve faster and more efficient outcomes.

UNDERSTANDING THE IMPACT

Agile is not solely a set of practices; it encompasses psychology, behavior, and cultural aspects as well. Merely implementing Agile practices without understanding their impact on the organization and its culture can lead to failure, with Agile being unjustly blamed. Therefore, it is crucial to recognize that Agile will affect every aspect of the organization, including contracts, hiring practices, progress measurement, reporting, and stakeholder engagement. The adoption of Agile practices and the promotion of Agile behaviors will have wide-ranging effects, which will be discussed in future sessions.

Furthermore, an Agile culture can benefit the organization even if traditional methods are still in use. The principles underlying Agile, such as hiring motivated individuals and focusing on delivering faster and doing the right thing, are valuable regardless of the specific practices employed. By embracing an Agile culture, organizations can expect improvements.

Leadership must also acknowledge that the change brought about by Agile will affect them as well. Some organizations mistakenly perceive Agile as limited to the teams, expecting everything to remain unchanged for them. However, it is crucial to understand that tracking progress and other aspects will inevitably change with Agile's implementation.

QUESTION 11

In Agile, which of the following is fixed?

A. Schedule and budget

B. Schedule

C. Budget

D. Scope

ANSWER

The correct answer would be scheduled and budget.

Because you have to have those two fixed points. We don't have to but they're strongly encouraged. The question is how? Right? How do you fix your schedule and budget? That sounds like it's really difficult but again, as a reminder, we like working in timeboxes, meaning define start, defined end. We're going to talk about sprints and iterations, which are a fixed timebox typically every two weeks, a release could be a fixed timebox of whatever is appropriate, and the project itself is a fixed timebox.

So you can fix the schedule. Now, if you fix the schedule and you have a stable team or collection of teams within that project then now you've got fixed budget as well, because it's the labor cost as you're typically your highest, not always, you know. Obviously, you might have some fixed assets and fixed things that are also part of that budget. But you can fix the schedule and the budget simply by stable teams and stable timebox. That means that the one variable that you can get away with is scope.

AGILE CHALLENGES

When adopting Agile, organizations commonly face several challenges and concerns. Let's examine them section by section:

1. Organizational culture: Many organizations struggle with aligning their existing culture with Agile concepts, such as decentralizing decision-making and allowing self-organizing teams. Embracing concepts like avoiding overtime and focusing on outcome over output can be difficult for some organizations.

2. Resistance to change: Some organizations claim that Agile cannot work for them without providing a clear reason. Agile's purpose is to change the way work is done, and it can be applied to various environments. Making Agile work requires a shift in understanding and thinking about work.

3. Lack of understanding: Training provides a better understanding of Agile, but the journey doesn't end there. Practicing and experimenting are crucial, even if failures occur. Learning from these failures and adjusting along the way is essential, and it's beneficial to fail early.

4. Tailoring too soon: Agile allows tailoring, but it is important not to tailor the process too soon. It is recommended to follow the practices initially, gain a deeper understanding, and recognize the value before making tailored adjustments. Tailoring should not be based solely on fitting the current culture but on the right reasons and the right way.

5. Agile in name only: Merely following Agile practices does not make an organization truly Agile. It is the behaviors, culture, and mindset that truly embody Agile. While following the practices is commendable, it is important to remember that Agile encompasses more.

CONCERNS AND FEARS

All right, let's move on and discuss some concerns:

1. Loss of control: The decentralization of decision-making and the allowance for changing requirements may initially feel like a loss of control. However, Agile, when implemented effectively, offers transparency, openness, and better control through informed decision-making.
2. Job role changes: Managers and project managers will experience changes in how they perform their roles. Understanding and adapting to these changes is crucial.
3. Poor quality: Concerns about a decline in quality are unfounded. While the way testing and validations are approached may change, customer satisfaction and high levels of quality remain top priorities.
4. Change chaos: Allowing change does not mean chaos. Agile has lightweight constraints and emphasizes frequent conversations about project vision, roadmaps, and scope. Change is controlled and prioritized.
5. No planning: Agile involves continuous planning at different levels, from vision-mission statements to daily planning. Planning is a significant aspect of Agile, often more extensive than in traditional methods.
6. Lack of productivity: As long as organizations have hired capable individuals and established transparency, concerns about people not doing their jobs should be

alleviated. Employees are motivated to perform their responsibilities, and it is essential for leaders to support and enable their work.

7. Lack of control over team decisions: Agile incorporates controls through the role of the product owner, who represents the scope and serves as a business representative. Transparency and visibility ensure that teams move in the right direction and adhere to the agreed-upon decisions.

By addressing these challenges and concerns, organizations can navigate the Agile transformation more effectively and reap the benefits of Agile practices and principles.

ROLES

Stakeholders are the starting point in the Agile process. They are the ones brimming with ideas and requirements for what needs to be built. Often, these ideas might originate from customers or other sources within the organization.

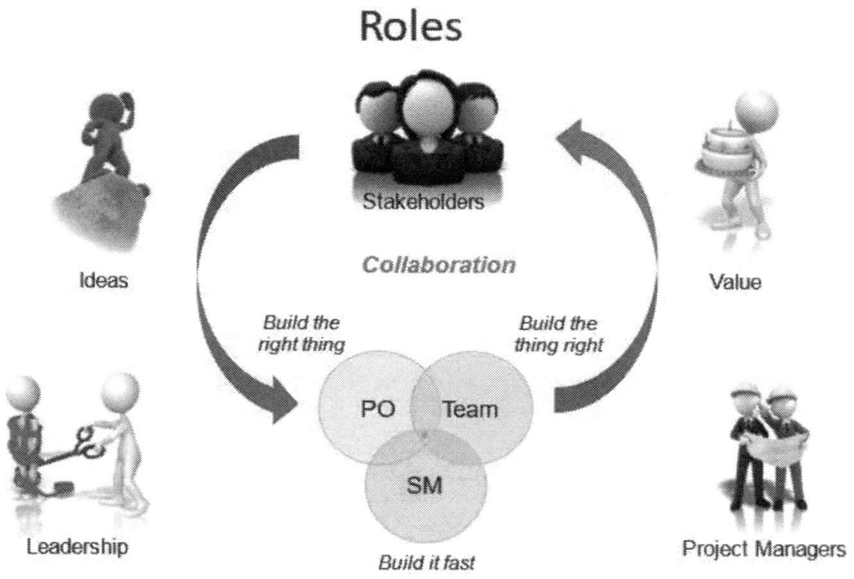

FIGURE 10: STAKEHOLDERS IN AGILE ENVIRONMENTS

Once these ideas are gathered, they are channeled to the team. This team comprises three primary roles:

- **Product Owner:** The product owner is the voice of the stakeholders within the team. They ensure that the team is focusing on building the right thing. Typically, this role is filled by someone from the business side or someone who represents the customers, ensuring that their requests and needs are adequately represented.
- **The Team:** This group is ideally co-located, but in today's digital age, they can also be virtually co-located. They are a self-organizing, cross-functional unit equipped with the right skills to deliver high-quality work. Their primary responsibility is to ensure that whatever is being built is done the right way. If there are gaps in their knowledge, they are proactive in seeking out learning opportunities and experimenting to fill those gaps.
- **Scrum Master:** While not all Agile processes have a Scrum Master, many do, especially Scrum. The Scrum Master's role is to ensure that best practices are followed and to streamline the process as much as possible. Their focus is on speed and efficiency.

These three roles work in harmony. The Scrum Master emphasizes speed, the product owner ensures the right things are being built, and the team ensures it's built correctly. Their collective aim is to strike a balance, ensuring that the right things are being built in the right way and as quickly as possible. It's a delicate balancing act, with each role pulling in its direction, but the goal is to find a harmonious center.

Beyond these core roles, there are others that play a crucial part in the Agile process:

- **Leadership:** Their primary responsibility is to remove any impediments or blockers that might be hindering the team. They set the direction and provide the necessary support to ensure the team can get there.
- **Project Managers:** Contrary to some misconceptions, project managers are still very much relevant in Agile. Their role is to foster collaboration across teams,

engage with stakeholders, ensure governance protocols are followed, and protect the organization. They also play a pivotal role in communicating the status of projects and managing complexity and risk.

The collective goal of all these roles is not just to deliver a project but to deliver value to the customers. Value, however, is a subjective term. It can vary from one person to another, from one project to another. The challenge is to define what value means in each context and ensure that the entire process is geared towards delivering that value.

In essence, the flow is straightforward: Ideas originate from stakeholders and customers. These ideas are then represented by the product owner, built by the team, and streamlined by the Scrum Master. Project managers and leadership play supportive roles, ensuring that the process runs smoothly, risks are managed, and value is delivered to the customers.

THE TEAM

In the Agile framework, the team is the heart of the process. It's not just a group of individuals working in silos; it's a cohesive unit working towards a common goal. Let's break down the composition and ethos of this team.

Composition of the Team:

- **Product Owner:** This individual represents the voice of the stakeholders and ensures that the team is always aligned with the business goals and customer needs.
- **Scrum Master:** The Scrum Master acts as a facilitator, ensuring that the team follows Agile best practices and remains efficient and effective in their tasks.
- **Developers or Development Team:** These are the individuals responsible for building the product. They possess diverse skills, ensuring that the team can handle various aspects of product development without external dependencies.

Both the Product Owner and Scrum Master are integral parts of the team, not external entities. They play crucial roles in guiding and supporting the team.

Characteristics of the Team:

- **Stable:** The team remains consistent, which helps in building rapport and understanding among members.
- **Cross-Functional:** Team members have varied skills, ensuring that they can handle different facets of product development.
- **Self-Organizing:** The team takes ownership of their tasks, organizes their work, and determines the best approach to achieve their goals.
- **Size:** Ideally, the team is small, typically comprising seven to nine members. A fun way to gauge the team's size is the "two-pizza rule," suggesting that the team should be small enough to be fed with two pizzas.

Team Ethos: The essence of an Agile team lies in its collaborative spirit. It's not about individual achievements but collective success. If a team member completes their tasks ahead of time, the expectation isn't for them to start something new indepencently. Instead, they should assist other team members, ensuring that the team's collective tasks move towards completion. This approach ensures that work is completed faster, and the team learns and grows together.

The phrase "that's not my job" has no place in an Agile environment. Every team member is focused on the team's overall goal, not just their individual tasks. The emphasis is on finishing ongoing tasks rather than starting new ones, aligning with the push versus pull philosophy. This collaborative approach ensures that the team delivers value quickly and efficiently, embodying the core principles of Agile.

QUESTION 11b

The Ag le Manifesto states that the sponsors, developers and users should be able to maintain ___________

A. A fast pace through the project
B. A constant pace indefinitely
C. A fast pace for part of the sprint
D. A quick rate of progress at project start

So the reason B is different from A, and that's perfectly understandable why people would put A, is that remember that we want to view it more as a product-focus than a project-focus. So we want to build these teams that can maintain their pace for as long as that team is in existence. Hopefully it's longer than a single project, because we are building this team and they're working really well together, let's not break that apart, because then you've just lost all the value of that team. Let's consider it, let's start thinking more about longer term rather than just that moment in time of a project. There should be a constant pace indefinitely, and hopefully a fast pace to go with it but faster doesn't always mean better. If we're going really fast and producing a bunch of garbage, that's not what we want to do. We want a constant pace, maintaining quality standards. Constant pace does not necessarily mean fast pace!

Answer: The correct answer is B, a constant pace indefinitely.

QUESTION 12

Simplicity – the art of _______________ is essential

A. Maximizing the amount of work done
B. Maximizing the amount of work not done
C. Maximizing the amount of features done
D. Maximizing the amount of money spent

I find this statement really funny in the manifesto. I have no idea why they worded this way but I heard a great quote, I forget who said it but they said "Given the opportunity people will find the most complex way of stating the simplest concepts" and this is a classic example of this. This whole thing is about simplicity, and they found the most complex way of describing simplicity, that's possible. So the correct answer is B, the art of maximizing the amount of work not done is essential. Basically what that means is find ways of not doing work, right? Because if you're not doing something, it frees up opportunities and time to do stuff that's actually more important. So surely, you can think of things in your lives, in your projects that are not really that necessary, maybe there's a way of getting rid of those, let's simplify our process, let's simplify the product. What it is that we're building and how we build it so that we can reduce the amount of work that we're doing.

Our goal is to is to do as little work as possible to produce the best outcome possible, all right? Think about it that way, do less to produce better outcomes, sounds backwards but it really works.

Be specific, so there's intra and inter-team relationships right? So inter meaning, across teams and outside the team. How do we how do we work together outside of our team.

Here's the thing, most training sessions are going to be team-focused. How things happen inside the team and how a team reacts. We also want to talk about how those teams how multiple teams can interact with each other, right? So we'll talk about a little bit in later session but I'll give you a little preview that now we're talking it to a certain level about scaling, how do we do more complex projects where a single team doesn't have the capacity to get it done within the amount of time that's required. Now you've had a complexity into your project and you need to put mechanisms in place so the teams talk with each other and stay in alignment, or you're going to increase the risk of your project and that's bad, bad stuff, right?

You have somebody on the team that is helping to manage scope and that person is in contact with the customer, so it's a kind of a joint discussion. We never want to do things just on the customers behalf without talking to them. So it falls down to collaboration and there may be some contractual things that we have to do as well but it is absolutely managed. It's not like we're just going to change for the sake of change.

Answer: Maximizing the amount of work not done

CHAPTER 3

SCRUM ESSENTIALS

SCRUM OVERVIEW

In this overview, we're going to be talking about some tangible concepts, and, as we work through the concepts and practices, things will start to click. Scrum can be described as the most widely used framework in agile. It has been said that out of the firms using agile, approximately 90% of them use Scrum or parts of Scrum. As documented in *The Agile Assessment Guide* (published by the U.S Government Accountability Office); the main purpose of scrum is to address complex problems while delivering high value products frequently. Scrum originated from a 1986 text by Hirotaka Takeuchi and Ikujiro Nonaka titled, "The New New Product Development Game," the framework was first referred to as "Scrum" by Ken Schwaber and Jeff Sutherland in the early 1990s to emphasize a holistic approach using multiple, overlapping phases.

Schwaber and Sutherland authored the Scrum Guide, which details the framework. Scrum relies heavily on the concept of "Scrum teams" that are responsible for producing working software in increments often referred to as a "sprint." Each sprint is a short, time boxed iteration that is intended to provide distinct, consistent, and incremental progress of prioritized software features. The Scrum framework is centered on Scrum teams where members fill specific roles and responsibilities. These members are responsible for various tasks, including developing Agile artifacts. Each team contains members that fit into one of these three main roles.

With Scrum, teams are self-organizing and choose how best to accomplish their work, rather than being directed by management. Teams are also cross-functional, meaning they include members who have the capabilities to achieve the work without depending on someone outside the team. This model optimizes flexibility, creativity, and productivity and seeks to eliminate the need for a traditional program manager since each team supervises itself.

During sprint planning meetings, the team determines the type of work to be done, prepares the sprint backlog (ordered list of tasks to be accomplished during the sprint), and communicates expected responsibilities between team members. Teams meet daily during each sprint for a brief status update. Each sprint is intended to produce, among other things, completed increments of software features that are ultimately built into the final product solution. The sprint backlog is a subset of the most important features from the overall product backlog. Teams decompose these requirements into user stories that describe what the customer wants. The software developed during the sprint should satisfy those needs in order for a user story to be considered complete. A burn-down chart is a public display of the remaining work in the sprint backlog. The team updates the burn-down chart daily to keep everyone informed of the status of tasks.

Diving Deeper into Scrum

To begin let's discuss the rule of 3-5-3, which are **3 Roles, 5 Events, and 3 Artifacts**. To be prepared for the discussion of these rules, you should have a Scrum Guide to use as a reference. This guide is free and can download in multiple languages by going to ScrumGuides.org.

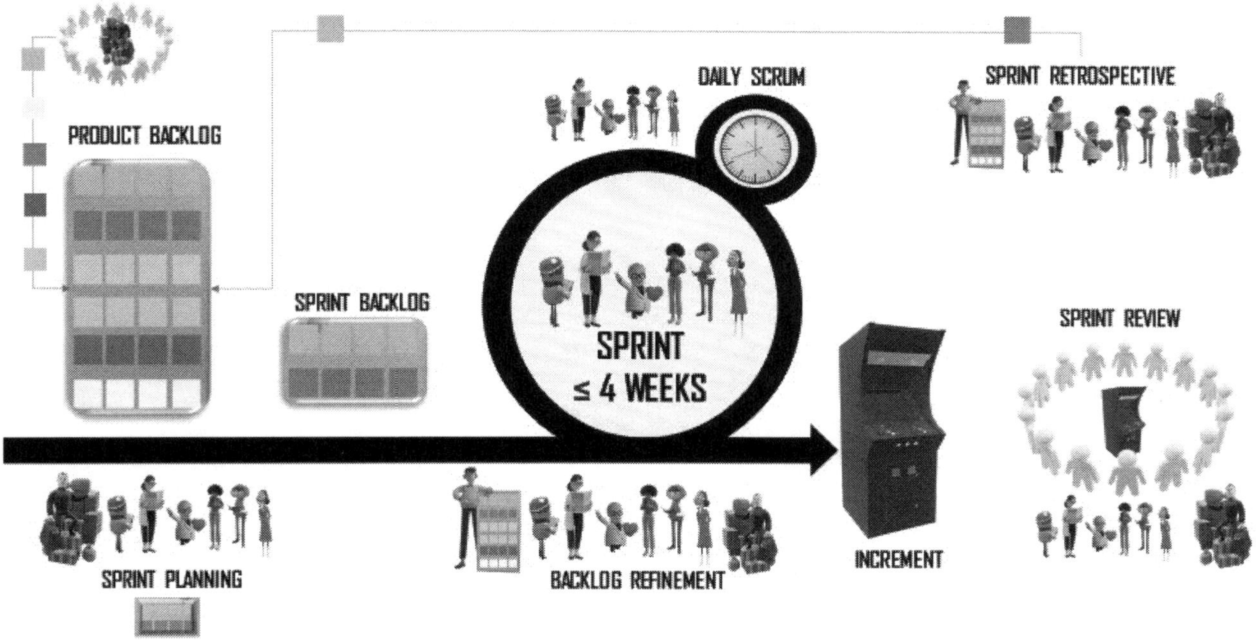

Figure 11: The Scrum Framework Summarized

An event is technically just a meeting. Most people dread meetings, so the purpose of calling it an event in Scrum makes it a little more palatable. When a meeting is mentioned, people generally get a feeling of dread in their stomachs. They feel as if they will be wasting precious time while there, and usually, they are just bored. We are endeavoring to change that negative mindset toward meetings. We want participants to feel that any Scrum related events are designed to bring value.

There are agendas to follow. There are techniques to follow, practices that teach as you go, and every one of them is designed to be valuable and have specific outcomes.

If you go into one of these meetings and you're not getting value from it, then we should take a very hard look at it and make the appropriate changes. You should never go into one of these sessions and say, "Yeah, that was just absolutely useless because it was done the wrong way." Now, you've probably gone to meetings where you just sit there, listening for an hour. For example, someone is presenting an Excel spreadsheet covering it line-by-line. We want to move away from those concepts to be more workshop-focused, more collaborative so that every meeting we attend has value, and if it's not, we want to remedy that by fixing it. That's one of our guiding concepts.

Let's start with the **3 Roles, which are the Product Owner, the Scrum Master, and the team**.

THE PRODUCT OWNER

The Product Owner owns what we are building. They are responsible for prioritizing the list of work, the backlog. We will be learning more about what a Product Backlog is and how to manage it as we move along. The Product Owner is also responsible for maintaining that content, and they are not able to do that job alone. To be extremely specific about this. Being a Product Owner is a hard job. It is a lot of work, and it's just almost impossible for one person to do it all. They are going to need help, which they receive from the Scrum Master, the team. Everyone can contribute to the Product

Backlog. These individuals work together collaboratively, but ultimately, the Product Owner is the final deciding voice. In every case, there needs that one person who says, "This is what we're going to do." The Product Owner must take all the information from stakeholders, from the team, and from the Scrum Master, and collectively come up with a decision about what is the right thing to do at that time That's the job with a Product Owner. It's not easy, so give them some space to do their jobs.

THE SCRUM MASTER

The Scrum Master is essentially your coach and facilitator. The person that knows the practices inside-out. They are, as this term says, Scrum Master, i.e., the master of scrum. They know it very, very well. And they're there to help set up the right guardrails so that we're good practices are being followed. The Scrum Master is there to make sure the team has everything they need. Or to look at it another way, they're there to make sure the team has the best day possible. Helping someone have the best day possible involves removing impediments, helping the team get through challenges, and that may involve just talking to them as a friend. Sometimes, someone is having a bad day. The Scrum Master can be there to say, "Hey, how is it going?" and just be aware of the well-being and the health of the team.

Keep in mind that a Scrum Master is different from a Project Manager. They are vastly different roles. They may potentially be the same person; however, their viewpoint and involvement are like two sides of a coin. Primarily, Project Managers are focused on the project – stakeholder engagement. They're focused outward from the team, making sure the community of stakeholders, the leadership, and other teams are given updates to keep them informed. The Project Manager makes sure that things are coordinated across teams. Essentially, they're focusing outward.

A Scrum Master focuses inward. They're focusing on the team themselves, not just about getting work done. Getting the work done is the main goal, of course, but the focus of the Scrum Master is making sure the team has what they need to get the job done. The Scrum Master is responsible for such things as making sure the team has the equipment, the training, and the knowledge they need to move smoothly through the completion of the project. To ensure this smooth flow, the Scrum Master removes all impediments that would interrupt the work. They are looking ahead and escalate when necessary to remove the stopgaps, etc. As previously mentioned, the Scrum Master not only handles all of the process kinds of things; they are also involved in the psychological and emotional well-being of the team. To do this, they must make sure team members are interacting well, collaborating properly, that they are managing conflict, thereby, eliminating any discord so the team can be a very high-functioning unit. There's a lot of focus on the behavioral aspect of the team, not just the team doing their work. That's a bit of a shift or might seem that way from how most processes operate. We bring the team into the process. Their emotional and psychological well-being is part of the process.

THE DEVELOPERS/TEAM

It is a proven fact that a happy team is a productive team. This group of individuals working together in harmony will produce much better stuff if they enjoy what they're doing. Therefore, it is essential to factor into the process of how to keep the team satisfied.

Now, the last role obviously is the team itself, and it is a cross-functional small self-organizing team, meaning they have the skill sets to do the job. They know how to do the job, and they can make decisions for themselves. This team's independence means you no longer have to worry about assigning work to the individual team members. Instead, they make those decisions themselves because they're smart people, and they can make those decisions.

Obviously, this decision-making process has to be scoped. It's not as if one day you're building a bicycle, and the next day you're building a house. Clearly, there are two drastically different things. Thus, it is going to be scoped with those decisions. The team is making a lot of decisions to accomplish a common goal. Hence, they need to make sure they are saying, "This is where we're going, and that's what your vision statement or your project charter is all about." "This is where we're going; this is what we're trying to build, and let's make our decisions along that path." By approaching decision-making from this aspect, it's not chaos as changes are frequently required; it's not "do anything you want to." It's focused along with the project goals, which is the concept.

Let's move on and talk about the **Artifacts** and the **Events.** Other things, in general, go on in Agile, and Scrum is just one of the many practices of Agile. It is the most widely adopted form of Agile. However, it doesn't mean that it's the right Agile practice. Don't just choose Scrum because everyone chooses Scrum. Instead, choose the practice that makes the most sense. However, since Scrum is the most widely adopted, we're going be talking about it the most in a few different practices throughout the course. Many concepts apply to all of the practices; it's not just Scrum.

Primarily, what we will be discussing are the activities that go on in the team. There are going to be likely many events that happen outside of the team, like making sure the team knows the right stuff to work on, making sure the program and portfolio activities happen as well. This is just the team focus. We'll talk about some of those other things a little bit later.

Let's start with a Product Backlog. The Product Backlog is the artifacts that include the Product Backlog, the Sprint Backlog, and the PSI (Potentially Shippable Increment.) As a side note, keep in mind that the PSI has multiple names. It could be potentially shippable product increment or Potentially Shippable Increment. There are many different names by which it could be called, but basically, it is a piece of the product that is potentially ready for use by your customer. This will also be discussed in detail later.

THE PRODUCT BACKLOG

A Product Backlog is a list of all of the things that we may want the team to work on. Using some "soft" language here only indicates the possibility of assignment to the team. Using words and phrases like 'may want' or 'should' or 'could' work on, does not mean we're going to do it simply because it's on the backlog. This goes back to that variable scope discussion. Items will show up on the backlog, but the decision will be made as a team along with our Product Owner and stakeholders as to whether it is the right thing to work on. If we decide it is not the right thing, it may be thrown out or changed to something more appropriate. Simply because it's in the backlog, doesn't mean we're going do it. It means likely we will do it.

So, the Product Backlog is a single prioritized list of everything that we want to work on or may want to work on. Now, it's really important to understand the Product Backlog is prioritized. Prioritized with the understanding that there's only one priority: just one, that's it. Not two priority ones, or any other number other than one. And that's extremely important for the team. The team is the wrong group of people to decide what's the most important thing to work on. They can help in the discussion of what the one priority should be. Their input and contributions are valuable because certain things need to be done in sequence to enable certain functionality or to ensure certain things happen. But when we're talking about the value of things, the business value, or customer value, they're not always the right people to make that decision. So, if they're not the right people to decide that, then you can't give them multiple priority ones. By doing so, they would be forced to decide what's the top priority, the thing that they need to work on first. Or worse, they'll time slice. They'll be juggling priorities and context switching, which reduces their effectiveness. To effectively help the team, be very aggressive about prioritizing the work that goes to them.

This is a core aspect of Agile. Prioritization moves significantly up on the list of things that need to be accomplished. This facet is hard, especially when you have multiple stakeholders, each one thinking that their stuff is the most important thing. So, remember the phrase: stop starting, start finishing. This is where it comes to fruition, right here. Prioritize your backlog. There is only one priority one, one priority two, etc. When we discuss this more, we give you some techniques on how to improve that process of prioritization, but it can be really difficult.

This list of things that we may want to work on could be requirements or features scope. It could be bugs, defects, things that we need to fix. The items could include operational things. Basically, anything that is going to take capacity from the team is on this list. Therefore, this is the reason we need to have a truly transparent view of what the team is working on. We need to see what their capacity is and whether they have the capacity to take on additional work. Absolute visibility is needed in everything that's going on, and that means if a team is working on multiple projects, you don't have visibility into just the projects. You have visibility of what the team is working on, which could be multiple projects. Hence, everything is visible on one board so we can see the work that's going on for that team. We'll talk more about storyboards as we get deeper.

THE SPRINT

The first day of beginning work is called the Sprint. The Sprint is an iteration. It's a time box. It starts on the same day and ends on the same day, every single time. A sprint may be 4 weeks or less, and that is not a variable. If you remember back in the manifesto, we want small deliveries, frequent deliveries so that we get more feedback. Thus, four weeks essentially only gives you 12 opportunities in a year to get feedback, while two weeks gives you 24 opportunities for feedback. So obviously, you're going to get more feedback, and that's better for everyone. However, this can also be hard as getting stuff done in 1 to 3 weeks may be very, very difficult. This will be a shift and something that we have to get our heads wrapped around. However, as you begin to think about things differently, you can do it.

SPRINT PLANNING

Once the sprint length has been selected, the very first day of our sprint, we do what's called a Sprint Planning event or a Sprint Planning event. This is where the team gets together with their Product Owner. The Product Owner or Scrum Master is considered part of the team. The team, the extended team with Product Owner and Scrum Master, get together, and they talk about all the things on the Product Backlog to determine what items should need to be worked on first. Now, keep in mind that the Product

Backlog is prioritized. Therefore, the discussion should be about items at the top of the list in a more detailed level of planning. These will be the things that we want to bring into our next sprit. In this discussion, all parties should be clearly understood, defining all requirements. Essentially what Sprint Planning does is it gives you a clear breakdown of the product backlog items (PBIs) to be performed during the sprint.

By the way, the industry average is a 2-week iteration. Also, just for clarification, sprint, and iteration are the exact same thing. It's just a time box. Few organizations do four weeks, but with the Scrum Guide saying a month or less, it is the standard used. One to 3 weeks is best, but here we are recommending two weeks.

The Sprint Planning session is essentially clear break-down of the work as detailed as the team needs it to be (sometimes down to the task-level). Through extensive and transparent breakdown. We know what the work is and who's going to do it. We know the hours required to complete the work. Essentially, we are building a work breakdown for the next two weeks. We don't build a work breakdown for the entire project, and the reason is that we want to allow change to happen very easily in a very cost-effective manner. If you're doing a WBS for your entire project, you're going to resist change because of the amount of time and effort that you spent building a WBS. So instead, we say we're going to compile high-level plans for what we think is going to happen. The high-level plan is followed by a slightly more detailed plan. Eventually, we get down into our sprint plan, which is incredibly detailed. We know everything about the work at that point. And we should have a rather good idea of what we're going to accomplish.

SPRINT BACKLOG

The Sprint Planning event creates The Sprint Backlog, and that is a small, more detailed list of work with all of the information that we need or the majority of the information that we need actually to get the work done. It is one of the 3 recognized Scrum artifacts. At the end of The Sprint Planning session, we sit down, we build our work breakdown structure, and the team agrees that yes, they feel confident that they can get this work done. It is always best to get that vote of confidence at the end where everyone says, "Yeah, we feel comfortable with this; we can do this work." If anyone expresses doubt indicating that they're not comfortable, then the team just keeps planning until everyone is comfortable. Otherwise, after the meeting is over, the team disburses to start work. This can lead to the work not getting done or other bad situations. This is not what you want to happen.

One of the values that we get out of The Sprint duration of 4 weeks or less is, firstly, we get a time box. That means that we know what should be completed within that specific amount of time. By the end of that sprint, we should expect that the work is completed. Potentially, there may be 1 or 2 things that don't get done. This is not a major catastrophe as the goal is still to get that work done.

The other thing that you get out of The Sprint is you get a sense of urgency. Instead of rushing to meet an upcoming deadline, or at the end of the project, we are working in a time-box to have things ready every two weeks.'

Imagine what that does to people who are procrastinators. The sense of urgency to get our jobs done is constantly there. By working within a two-week deadline, it encourages people to get their jobs done and not put things off until it's too late. That's another value that we get because urgency is incredibly useful.

After we've executed our Sprint Planning session, the work begins. We start executing and building things while continuing to work together. We're collaborating as a team to discuss the design and any problems encountered. No one on the team should isolate themselves to start building only to show up two weeks later. That's not our goal.

DAILY SCRUM

The other thing that we do as we're executing our sprint is, we have a Daily Scrum. The Daily Scrum is another event (one of the five). This event used to be called the Daily Standup but has become known as the Daily Scrum. The purpose of the Daily Scrum is a planning meeting. Experience has shown that this is the meeting that gets abused the most. The Daily Scrum can quickly turn into a status meeting that does not exist in any Agile practice. It just not necessary. The goal is that we pull status instead of push status. Therefore, status meetings are not needed. Anyone who needs status can see it in the Sprint Review event and other events as well that show the work happening. Status meetings get eliminated or reduced significantly. There's truly little value in them, so you should get rid of status meetings.

At the Daily Scrum, we talk about what we're working on and plan how we're going to work together as a team over the next 24 hours. This planning is the real purpose of the Daily Scrum, a planning event of how we're going to work together. This is a good opportunity for anyone to ask for help or discuss any needs or impediments to completing their part. There are a couple of different ways of running them, and we will talk about them in the future. For now, just consider the Daily Scrum is that daily sync point and the daily planning event to help the team plan those little tiny things that happen over the course of a sprint.

SPRINT REVIEW

The next event occurs at the end of The Sprint, and that's the Sprint Review. The Sprint Review is also called a "demo". This is an opportunity for us to invite our stakeholders, other teams, and people that are interested in what we're doing. It is a chance to show them what we did during the Sprint.

When you are first starting, finding something to demo can be difficult. At that point, you may not have a lot to show, so your first few Sprint Reviews may seem awkward. You will learn as you go through more of these reviews on how to come up with something to show the stakeholders.

The Sprint Review's main purpose is to get feedback. We show what we've built. Remember, the primary measure of progress is the working product. This is your opportunity to show the working product. Therefore, when your stakeholders come in, you can say, "This is what we just built over the past two weeks. What do you think?" This is tangible. It's something physical hopefully, or at least; it's something that we can show that we got something done. So think about it this way, your stakeholders are going to see something quantifiable, the promise that originally stated. This is what we're going to build, and, at the end of it, we're going to say here it is. Not we're 20% or 30% through, but here's the thing that we did. This aspect builds trust; it builds better progress and transparency. Your stakeholders can then start to back away, and leave you to do your work, because, based on what they've seen, they trust that you're actually doing your job. That's really important.

Another aspect of the Sprint Review is that it will generate ideas. Once the stakeholders or other members of the team see what you have presented, they may say, "Hey, that's cool, but can we do this?" What you have demoed to them has sparked additional ideas to build possibly. A point here to keep in mind is just because a stakeholder asks, "Can we do this?" doesn't mean that we're going to incorporate their ideas. Or possibly, their suggestions could be incorporated at a later date. A typical response to this question would be, "Hey, great idea! Let's put it on our backlog and feed it through our process." This enables the team to make sure that it's the right thing to do, and it's the right time to do it. If it's not the right thing to do, we're not going to do it. It's just that simple. Therefore, accepting the ideas doesn't mean you're going to do them.

SPRINT RETROSPECTIVE

The last event that we have is the Sprint Retrospective, and The Sprint Retrospective is your primary inspect and adapt mechanism. It's your opportunity to get better. If you have one meeting, the most important meeting in the entire series of events is the Retrospective. Never skip the Retrospective.

The results of all this are the Potentially Shippable Increment. It's a piece of the product that contain enough value to deliver it to our customers. Of course, that does mean you have to ship it to your customers. Two weeks may not be enough time to give anything of value to your customers. It might be months before there's something ready for them. What you're building is in increments, building it piece by piece by piece so that you have something tangible to show for progress and to get feedback as you continue working. That's the primary goal.

In some cases, for example, if you're building software, a Potentially Shippable Increment could possibly go into production, which is great. But if you're building something physical, the likelihood is that's not going to happen. Don't worry about it. You don't have to ship something from every sprint.

BACKLOG REFINEMENT

One last event is the Refinement event. The Backlog Refinement. This is an optional event. It's not in the Scrum Guide, but every team holds this event in some way. It’s somewhat built-in. That's why it's not one of the primary events. The primary reason for Refinement Event is to look at the work that's coming up in the future sprints, at least the next sprint, and talk about that work and refine it, improve it. By the time we get to our next Sprint Planning event, we understand the work to the point where we feel confident that we can get it done. The Refinement Event helps to improve your requirements along the way. So, instead of defining all the requirements upfront, you work on some of the requirements, and then improve them throughout the course of your project. That concept might take time for you to get your head wrapped around because it's not what you're used to. However, the result is we get much better requirements because we're taking the lessons that we've gained from each of the iterations, and we are applying it to the next set of requirements. Meaning the requirements are continually improving along the way rather than just defining everything upfront without the appropriate knowledge. Thus, this is one of our goals.

So, that's a basic overview. It sounds a lot easier than it really is. It's actually pretty hard, but we'll talk more about that soon.

To be specific, if you read the Scrum Guide regarding The Sprint Planning Event, it calls for a 2-week sprint whereby you should expect a 4-hour event. Experience shows that it is rare for it to go that long. Generally, an hour per week of the sprint, so if you're making a one-week sprint, typically an hour is good, with a 2-week sprint typically 2 hours is good.

1. **Sprint Planning Event:**
    - **Duration:** The Scrum Guide suggests a 4-hour event for a 2-week sprint. However, in practice, the duration varies based on the team's experience and familiarity with the process. Some seasoned teams can wrap up Sprint Planning in as little as 30 minutes for a 2-week sprint. The key is to ensure that the team understands the work, identifies risks, and maps out dependencies.
    - **Vote of Confidence:** After discussing the work, the team should collectively decide if they can confidently complete the chosen items within the Sprint.
2. **The Sprint:**
    - **Consistency:** Sprints start and end on the same day, ensuring a consistent rhythm. For instance, if a sprint starts on a Monday, it will end on a Friday two weeks later, with the next sprint starting the following Monday.
    - **Flexibility:** Some teams opt to start their sprints mid-week to avoid common days off like Mondays and Fridays. This ensures better participation during crucial sprint days.
3. **Product Backlog vs. Project Backlog:**
    - **Product-Centric:** The term "Product Backlog" emphasizes a focus on the product's entire lifecycle, not just a specific project phase. This approach encourages teams to think about the long-term value and evolution of the product.
    - **Dynamic Nature:** The Product Backlog is a living document, constantly evolving based on feedback, new requirements, and changing business needs. It can encompass items from multiple projects, reflecting the broader vision for the product.

4. Daily Scrum:

- **Potential Exclusions:** Some teams choose to skip the Daily Scrum on the first and last days of the Sprint. The rationale is that the first day is dominated by Sprint Planning discussions, and the last day focuses on the Sprint Review and Retrospective.
- **Purpose:** The Daily Scrum is not just a status update meeting. It's an opportunity for the team to synchronize, discuss challenges, and plan for the day ahead. If the team finds value in it, even on the first and last days, they should continue with it.

5. Backlog Refinement:

- **Distinct Role:** Backlog Refinement is a separate event where the team reviews and refines the Product Backlog items for future sprints. It doesn't replace any other Scrum events.

In Summary:

Scrum events are designed to foster collaboration, ensure alignment, and drive value delivery. While the Scrum Guide provides a framework, teams should feel empowered to adapt the events based on their unique needs, always keeping the core principles of Scrum in mind. The ultimate goal is to deliver value to the customer efficiently and effectively.

QUESTION 13

A sprint is a ______

A. Time box

B. Scope box

C. Cost box

D. Risk box

Now, you may consider that other things are somewhat time box or box within that sprint like scope is definitely boxed. We tend to resist change within The Sprint. Meaning we don't want things coming into The Sprint too much because it destroys the sprint plan. That's the only time that we tend to be a little sensitive about scope change. But also, if you think about it, it's kind of a cost box, too. The correct answer is time box, but it also does kind of box-out the cost a little bit, too, if you think about it. If it's a fixed team, you know, fixed hours, fixed time, then that kind of fixes the cost. So that's one of the ways that we get that fixed point of cost as well.

Answer: Time box

QUESTION 14

The usual recommended duration of a sprint is typically ________

A. 2 weeks

B. 1 - 3 months

C. 3 - 10 weeks

D. 2 months

The correct answer could be 1 of 2 possible answers here, depending on how you view it. The answer, according to the Scrum Guide, is actually 1 to 4 weeks. The usual recommended duration for a sprint is two weeks. That's what most organizations select as their sprint duration timeframe. We want shorter time boxes because, if the span of the sprint duration is too long, you lose the opportunity for feedback. This is such a critical aspect of making sure you are going in the right direction, show it to your customers, talk to your customers. Without that feedback, you could execute for three months and have a monumental failure. That's a very costly failure at that point, so shorter iterations mean less costly failures. If you consider each one of these sprints as an experiment, you might go the wrong direction. Hence, keep them short to save costs and to help prevent the opportunity for failure.

Answer: 2 Weeks

QUESTION 15

The Scrum Master has authority over the team and is like a project manager in a strong matrix.

A. True

B. False

In Agile, we take a hugely different approach. Instead of having authority over the team, we want the team to have authority over themselves. Therefore, in this case, a Project Manager or Scrum Master does not have authority over the team. In fact, they are just considered another team member. They are not managers. Hence, the term scrum master and not scrum manager. Although there are some cultures where we do say scrum manager, the Scrum Master is a facilitator. They're that servant leader, that person there to help the team, but they do not have the authority to tell people what to do. That's the team's job to do it themselves.

Answer: The correct answer is B. False.

Something to think about, if you are a Project Manager and if you would like to become a Scrum Master, I encourage you to do so at the very least to understand it better. But also, keep in mind that the mindset of project management is different from the mindset of Scrum Master. It's a different perspective, so try to make that context switch, that mindset shift first. Otherwise, what you're going to do is you're going to be project managing a team. In this role, if you follow the same mindset of project, you're going to become more detail-focused, much more telling people what to do, things like that because that's part of the project manager's job. Scrum Master is the opposite. We don't tell people what to do. We coach them. We facilitate, we encourage them, we mentor them, and it's more like the Scrum Master is pulling the team along and encouraging them to do things and helping them discover the right answers rather than telling them all the time. It's a very different mindset from project management.

SCRUM VALUES

Scrum, as a subset of Agile, not only provides a structured approach to product development but also emphasizes the importance of certain values that drive behavior. These values are the bedrock upon which Scrum practices are built, ensuring that teams not only follow processes but also embody the spirit of Agile.

1. Courage:

- **Definition:** Courage in the Scrum context means having the bravery to do what's right for the product and the team, even if it's against the prevailing opinion or seems challenging.
- **Application:** Courage is manifested when team members voice their concerns, challenge the status quo, or advocate for a course of action they believe in. It's about standing up for one's convictions and being willing to take risks for the betterment of the product.

2. Focus:

- **Definition:** Focus is about channeling energy and efforts towards the most important tasks that deliver value.
- **Application:** Teams demonstrate focus when they prioritize tasks, avoid multitasking, and ensure that they're working on items that align with the product's goals and vision.

3. Commitment:

- **Definition:** Commitment is about being dedicated to the team's goals and the tasks at hand. It's a promise that team members make to each other to deliver their best work.
- **Application:** Teams show commitment by setting realistic goals during Sprint Planning and then working diligently to meet them. It's about being accountable and owning one's responsibilities.

4. Respect:

- **Definition:** Respect is about valuing the contributions of every team member and treating each other with dignity and consideration.
- **Application:** Teams practice respect by listening to each other's opinions, providing constructive feedback, and recognizing the strengths and skills of each member.

5. Openness:

- **Definition:** Openness is about being transparent in actions, challenges, and progress. It's about creating an environment where information flows freely.
- **Application:** Teams embody openness by sharing their progress in Daily Scrums, being honest about challenges, and being receptive to feedback.

The Power of Values:

Values, as Roy Disney aptly put, are the compass that guides decision-making. In the world of Scrum, these values ensure that teams don't just go through the motions but truly embrace the Agile mindset. They act as a beacon, guiding teams through challenges, ensuring that they remain aligned with the Agile principles, and fostering an environment of trust, collaboration, and continuous improvement. By internalizing and practicing these values, Scrum teams can navigate the complexities of product development with clarity and purpose.

In the realm of Scrum, two values stand out when it comes to fostering a healthy team environment: Openness and Respect. These values are intertwined, each reinforcing the other, and together they create a foundation for effective communication, problem-solving, and collaboration.

1. Openness: The Key to Identifying and Addressing Issues

- **Definition:** Openness is about being transparent in all team activities, challenges, and progress. It's about creating an environment where information flows freely, and issues are brought to light without fear.
- **Application:**
    - Teams should be encouraged to share their progress, challenges, and blockers in daily stand-ups and retrospectives.
    - Mistakes should be viewed as learning opportunities, not failures.
    - Teams should feel safe to voice concerns, offer suggestions, and ask for help when needed.

2. Respect: The Foundation of Trust and Collaboration

- **Definition:** Respect is about valuing each team member's contributions and treating everyone with dignity and consideration. It's about creating an environment where everyone feels valued and heard.
- **Application:**
    - Team members should listen actively to each other, avoid interrupting, and provide constructive feedback.
    - Differences in opinions should be addressed with understanding and empathy.

- Recognizing and appreciating each member's strengths and contributions fosters a positive team culture.

The Interplay Between Openness and Respect:

Openness and Respect go hand in hand. For a team to be truly open, there must be an underlying foundation of respect. When team members feel respected, they are more likely to be open about their challenges, ideas, and concerns. Conversely, when there's a culture of openness, it fosters mutual respect as team members see the value of each other's contributions and perspectives.

Challenges and Solutions:

- **Challenge:** Fear of retribution or blame can hinder openness.
    - **Solution:** Cultivate a blame-free environment where mistakes are viewed as opportunities to learn and grow.
- **Challenge:** Differences in opinions can lead to conflicts.
    - **Solution:** Encourage constructive discussions and focus on finding common ground. Respect differing viewpoints.
- **Challenge:** Management or stakeholders might react negatively to issues being highlighted.
    - **Solution:** Educate stakeholders about the values of Scrum and the importance of addressing issues early. Emphasize the long-term benefits of transparency.

Conclusion:

In Scrum, the values of Openness and Respect are not just words on paper; they are essential practices that drive team success. By fostering an environment where team members can be open about their challenges and are treated with respect, teams can navigate the complexities of product development with clarity, collaboration, and a focus on continuous improvement.

CHAPTER 4

THE 3-5-3 OF SCRUM

In this section, we will revisit all the roles, events and artifacts of scrum, going a lot deeper into specifics. Let us begin with the product owner.

THE PRODUCT OWNER

The Product Owner (PO) is a pivotal role in the Scrum framework. They act as the bridge between the business and the development team, ensuring that the product being developed aligns with the business's goals and customer needs.

1. Who is the Product Owner?

- **Representation:** The PO represents the voice of the customer. They should ideally be someone from the business side who understands the customer's needs and can articulate them to the development team.
- **Decision-Maker:** The PO has the authority to make decisions about the product, from feature prioritization to release timelines. They need to be empowered to make these decisions without constant oversight.

2. Responsibilities of the Product Owner:

- **Backlog Management:** The PO is responsible for maintaining the product backlog, which is a prioritized list of features, enhancements, and bug fixes for the product.
- **Stakeholder Collaboration:** The PO collaborates with stakeholders to gather requirements and feedback, ensuring that the product aligns with business goals.
- **Team Collaboration:** The PO works closely with the development team to clarify requirements, answer questions, and provide feedback on the product.

3. Challenges Faced by the Product Owner:

- **Time Commitment:** Being a PO is not a part-time job. They need to be available to the team, stakeholders, and customers.
- **Balancing Act:** The PO often has to balance competing demands from different stakeholders and prioritize features that deliver the most value.
- **Decision Making:** The PO must make tough decisions, sometimes with incomplete information. They need to be confident in their choices and be able to justify them.

4. The Value of a Good Product Owner:

- **Building Trust:** A good PO fosters trust between the business and the development team. They ensure that both sides understand each other's challenges and work together towards a common goal.
- **Efficient Development:** With a clear vision and well-defined requirements, the development team can work more efficiently and avoid costly rework.
- **Delivering Value:** The PO ensures that the team is always working on the most valuable features, ensuring that the product delivers maximum value to the customers.

5. Conclusion:

The Product Owner plays a crucial role in the success of a Scrum project. They ensure that the product being developed aligns with the business's goals and meets the customers' needs. By fostering collaboration, making informed decisions, and prioritizing effectively, the PO helps the team deliver a successful product.

THE SCRUM MASTER

The Scrum Master (SM) is a pivotal figure in the Scrum framework. Unlike traditional roles, the Scrum Master is not an authoritative figure but a servant-leader, ensuring that the Scrum team adheres to Scrum practices and principles.

1. Role of the Scrum Master:

- **Coach and Mentor:** The SM acts as a coach, guiding the team towards self-organization and high performance. They mentor team members, ensuring they understand and follow Scrum principles.
- **Impediment Remover:** The SM identifies and removes obstacles that might hinder the team's progress. This could range from organizational blockers to personal conflicts within the team.
- **Facilitator:** The SM ensures that Scrum events (like Daily Stand-ups, Sprint Planning, and Retrospectives) are effective and productive.

2. The Scrum Master's Relationship with the Team:

- **Peer, Not a Boss:** The SM is seen as a peer, not as an authoritative figure. This fosters an environment of trust and collaboration.
- **Empowerment:** The SM empowers team members, encouraging them to make decisions and take ownership of their work.
- **Continuous Improvement:** The SM promotes a culture of continuous improvement, encouraging the team to reflect on their processes and find ways to be more efficient.

3. Challenges and Considerations:

- **Balancing Act:** The SM must strike a balance between guiding the team and allowing them to make decisions independently.
- **Avoiding the Halo Effect:** The team should not rely solely on the SM for decisions. The wisdom of the entire team is more valuable than the opinion of one.
- **Management and Authority:** The SM should not be seen as a manager. Their role is to serve and support, not to command.

4. The Value of an Effective Scrum Master:

- **Building Trust:** An effective SM fosters trust within the team, creating an environment where team members feel safe to express their opinions and concerns.
- **Promoting Collaboration:** The SM encourages open communication and collaboration, ensuring that every team member's voice is heard.
- **Driving Results:** By ensuring that the team adheres to Scrum practices, the SM helps in delivering high-quality products on time.

5. Conclusion:

The Scrum Master plays a crucial role in the success of a Scrum team. They ensure that the team follows Scrum practices, removes obstacles, and fosters an environment of trust and collaboration. By serving the team and promoting Scrum values, the Scrum Master helps the team achieve its goals and deliver value to the stakeholders.

THE SPRINT

Let's explore Scrum events, beginning with The Sprint. The Sprint has a set duration, that could be 4 weeks or less, and this duration remains constant. Now, one might wonder: why isn't the duration adaptable? Why not alter it between one, two, or three weeks based on the nature of the work? This is a common query. Should we adjust The Sprint's length according to the work? The answer is no. It's crucial to remember that we set a fixed time and budget while allowing the scope to vary. Keeping a consistent time box also ensures a consistent schedule. But what if, nearing the end of that time frame, we haven't completed all we set out to do? The duration remains unchanged. The Sprint concludes as initially planned. However, at its close, we evaluate the circumstances.

What measures can we adopt to ensure task completion? Unfinished tasks are generally transferred to a subsequent sprint but it may not be the very next sprint. In practice, if a task remains undone, it's reverted to the backlog. Here, its significance is reassessed. If deemed crucial, it's allocated to the next sprint. Work progresses and tasks are eventually completed. Yet, such instances grant us a chance to scrutinize our planning method, refining it for enhanced consistency in subsequent sprints. This structure is purposefully designed to stimulate challenging dialogues – exactly what's needed. Identifying and addressing issues is possible only when they're recognized, deliberated upon, and collaboratively resolved.

When collaborating with companies new to Agile, a recurrent grievance I encounter is, "This Agile approach is spawning myriad issues. It simply isn't aligning with our operations." My response, consistently, is, "Actually, Agile isn't the root of your challenges; it's merely highlighting them." These issues have likely persisted unnoticed for a while. However, with Agile, turning a blind eye is no longer feasible. These challenges demand rectification. If left unresolved, they'll persistently rear their head, thwarting progress — a situation nobody desires. Ignoring issues doesn't lead to their resolution; proactive measures do. The beauty of Agile lies in its transparency. It spotlights existing challenges, prompting remedial action. Failing to address them ensures their recurrence, an endless loop of stagnation.

Let's discuss Sprint Planning, a pivotal event designed specifically to determine tasks for the upcoming time box. Typically, the recommended duration for this is 1 to 2 hours per week of sprint. A common misconception I encounter is the assumption that shorter sprints, say of a week, lead to excessive meetings, hence consuming more time. Contrarily, the thought is that longer, 4-week sprints reduce meeting durations. This is a fallacy. The duration of these events is proportionate to the sprint's length. For instance, meetings in a one-week sprint will be more concise compared to those in a 2, 3, or 4-week sprint. So, it's a myth that longer sprints equate to fewer meetings. In reality, meetings scale up with extended sprint durations. With a one-week sprint, Sprint Planning might last an hour or even less. Regardless of the sprint duration, the daily standups remain a consistent 15 minutes. Plus, shorter sprints entail briefer reviews and retrospectives. It's a misconception to assume longer sprints reduce meeting durations. Meetings adapt based on sprint length.

The primary objective of the Sprint Planning session is to reach a consensus on upcoming tasks, meticulously planning and ensuring team commitment to the sprint's goals. While the Scrum Guide doesn't use the term "commitment", I'm inclined to be traditional here. Team commitment can be as straightforward as the team expressing confidence in the sprint's agenda, fostering assurance among all participants.

However, the reluctance within the Agile community to use the term "commitment" stems from its rigidity. Committing to a specific plan or timeline might deter adaptability. Though we promote adaptability in Agile, given that we're planning merely two weeks of work, there should be a moderate level of commitment, ensuring a balance between adaptability and dedication, making it more feasible to fulfill the work planned.

Sprint Planning necessitates everyone's attendance. This isn't a mere formality; it's a call for collaboration and collective ownership. When a team crafts its plan, it doesn't merely allocate tasks; it instills responsibility. Contrarily, if tasks are simply thrust upon individuals with dictated timelines, ownership and accountability wane. It's like telling someone, "This is yours," as opposed to letting them claim it. When a team actively chooses its tasks, it fosters personal responsibility and fortifies team cohesion. Thus, universal participation isn't optional; it's essential.

Moreover, a pre-ordered and assessed backlog is pivotal. There's a prevalent misconception suggesting backlogs are only estimated on The Sprint's inaugural day. I contest this approach vehemently. A proactive assessment, prior to the sprint's onset, is vital for effective forecasting. Overlooking this becomes a narrow, team-focused perspective, betraying a naïveté within the Agile approach. For a clear forward vision, estimations in anticipation of The Sprint are non-negotiable. While there are nuances to estimating, which we'll address subsequently, the underpinning principle remains paramount.

Though release planning is discretionary, its merits shouldn't be underestimated. The plan's utility will be expounded upon later. Furthermore, the presence of a Product Owner is essential. As the conduit between the team and the backlog, they answer arising queries, clarifying the minutiae. Similarly, Subject Matter Experts are crucial, ensuring the team's decisions are well-informed.

Occasionally, knowledge gaps may arise; strategies to navigate these will differ. As for execution, it's relatively linear: prioritize backlog items, demarcate specific tasks, and gauge team capacity. This capacity evaluation, undertaken every sprint, contrasts available hours with required hours for task completion, ensuring realistic commitments. While seasoned teams might eschew this due to their adeptness, it remains invaluable, especially for novices. It's a safeguard against overreaching, a way to ground planning in pragmatism. Commitments should reflect genuine capability, perhaps nudging the boundaries of comfort, but never crossing into overextension.

In every sprint, we assess our accomplishments using a metric known as 'velocity', a pivotal measure in Scrum. This metric encapsulates our productivity. When we embark on sprint planning, we refer to our average velocity to guide our commitments. For instance, if our velocity indicates we typically complete 20 tasks (given a specific weighted estimate), we'd aim to commit to a similar amount.

The art of estimating in Agile differs slightly, which we'll venture into later. If past data shows we can complete 20 tasks, but we ambitiously commit to 40, we're setting ourselves up for potential failure. Such overcommitment isn't productive. Instead, realistic planning, aligned with our demonstrated capability, ensures sustainable productivity. It refrains from overburdening the team and aids in more accurate forecasting. Commit to what's achievable, ensuring both sustainability and predictability.

DAILY SCRUM

Let's dive into the Daily Scrum, followed by QA discussions. The Daily Scrum is a concise, 15-minute sync-up held daily. Yes, as the name suggests, it's daily. It's baffling how often I've heard, "We do our daily standups about twice a week." That's not how it functions. The term 'daily' denotes its regularity. This routine session is a chance for the team to strategize and highlight any hurdles they face. Given the limited 15-minute window, it's mainly about flagging issues rather than resolving them. We often follow the standup with a separate breakout session for in-depth problem-solving discussions, keeping the original meeting short and to the point.

I'm particular about punctuality for daily standups. They start and end as scheduled, emphasizing efficiency. If standups are consistently running long, we need to reassess our approach. Ideally, it's a rapid 15-minute chat about the tasks, potential impediments, and then it concludes.

We'll venture into varied techniques of running this session shortly. However, some ground rules are:

1. **Consistency:** The meeting occurs at the same location and time daily.
2. **Punctuality:** Everyone agrees on the timing, so be prompt. If scheduling is an issue, perhaps start five minutes past the hour, allowing for transitions. Absences should be exceptions, not norms, and in such cases, update the team beforehand.
3. **Preparation:** Arrive ready, limiting distractions. Remote participants will, of course, need their devices.
4. **Team Focus:** This is a team-centric session, not for management updates or status reports. We encourage managers to avoid attending, ensuring team members communicate among themselves.

Concerning execution, there are primarily two methods I advocate:

1. **Three Questions Approach:** Participants answer:
    - What did I achieve yesterday to move us towards the sprint goal?
    - What will I do today to move us towards the sprint goal?
    - Are there any obstacles or impediments in your way?

While these questions aid planning, they can become monotonous, risking the meeting's devolution into mere status updates. Beware it does not turn into this anti-pattern!

2. **Walk the Board:** This approach involves a visual board detailing tasks. Teams discuss tasks from the rightmost column (those nearing completion), focusing on steps needed to progress them further. Instead of a generic status update, the emphasis is on actionable steps to move tasks forward.

Both techniques have their merits, and it's essential to choose based on team dynamics and preferences. We'll explore these in more detail, especially with visual aids like storyboards.

QUESTION 16

The Scrum Master focuses on __________

A. Team well-being

B. Schedule and cost

C. Scope and risk

D. Management and authority

You might ask yourself what a Project Manager does? Project Managers typically focus on things like scheduling costs. Team well-being is the correct answer. That's what the Scrum Masters are primarily focused on. The last thing is team satisfaction. The Scrum Master should be trying to make that team happy, make sure they have the best day possible. If you're doing scrum well, your team is happy. They're enjoying what they're doing, and they're going to do it even better as a result. That's one of the primary purposes of the Scrum Master.

Answer: Team well-being

I always recommend doing something fun in your daily standup, bring a bad joke or you know, do something fun, sing a song, whatever. Just do something entertaining and socialize.

QUESTION 17

A sprint planning event will produce a __________

A. A product backlog

B. A sprint backlog

C. A project schedule

D. A project backlog

The term says it; Sprint Planning is to plan The Sprint, so that generates your Sprint Backlog. A Product Backlog is created over time, and it is the overall list of work that we need to accomplish for one or more projects but specifically for the product that we're building. A project schedule might be a piece of the Product Backlog. We tend to focus a little bit less on project schedules, although we do have something very similar. We don't get away from schedules. We can forecast, but it's a little bit lighter weight.

The correct answer is Sprint Backlog.

QUESTION 18

In Scrum, a collection of all User Stories is known as _______

A. A product backlog
B. A sprint backlog
C. A project schedule
D. A project backlog

This might be a little difficult because we didn't actually define User Stories yet.

Actually, User Stories will be covered soon, and we'll get to them very shortly. So yes, absolutely, technically, it could be A and B. A is the Product Backlog that is the correct answer; that is a collection of our User Stories and other things as well. All the stuff that we want to work on or might want to work on, the Sprint Backlog is really a subset of that. It's a smaller collection of stories that we're going to work on. The project schedule is definitely not because the schedule is more of a timeline, and User Stories are a form of a requirement. Project Backlog? There's no such thing.

QUESTION 19

Every day the team will meet at the start of the day (preferably) for a ___________

A. 15 minute daily stand-up meeting

B. 10 minute daily happy-hour meeting

C. 15 minute daily status meeting

D. 15 minute daily stakeholder reporting meeting

The correct answer is A. It's a 15-minute daily standup or Daily Scrum meeting. And the reason we call it a standup is that we do prefer that people stand up so that we stay focused, but it's not required. I use a little bit of old school with these things. I do encourage my teams to stand up, and I get lots of complaints about it.

Let me go to B for a second, so 10-minutes obviously is the indicator there, but it's not B. I have had people call them sit-down meetings. And I had one team that had someone who was confined to a wheelchair so to be inclusive of the individual on the team; they said let's have a sit-down, and they had some fun with it. It went very well for them. That was very creative for them to do that and again very, very inclusive, but definitely not a status meeting, because we don't have status meetings and that might feel a little weird. We don't actually need status meetings. The closest thing that we have to it is a review. We review what we accomplished over the past two weeks.

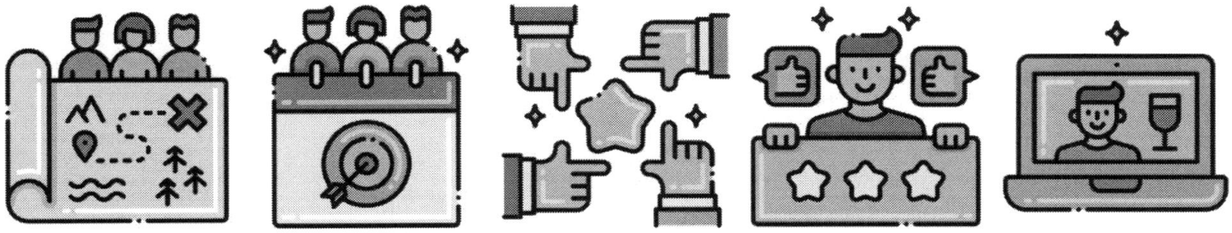

As you're transforming and becoming more and more Agile, you will likely have to do stakeholder readouts and things like that, at least for a while. Over time you can hopefully reduce them, and instead of having those traditional meetings, we can start bringing people into our new set of meetings and get better outcomes. However, just expect the daily stakeholder meetings to be there for a little while. Hopefully, we can get rid of them over time, but definitely not in a day!

BACKLOG REFINEMENT

Let's take a moment to learn about Backlog Refinement. Though not explicitly detailed in the core Scrum events or even the Scrum Guide, almost every Agile team incorporates this step. Its primary purpose is to clarify and fine-tune requirements. When it's time for your sprint execution, it's pivotal that tasks are well-understood, ensuring a smoother workflow.

Neglecting Backlog Refinement is a risky maneuver. Without refining or enhancing the understanding of tasks in the backlog, teams may find themselves in the Sprint Planning phase with a vague understanding of their responsibilities. This oversight can lead to numerous issues during execution, including unfinished tasks and decreased predictability. Such setbacks not only hamper the team's efficiency but also dampen morale, as repeated failures weigh heavily on team spirit.

Consider Backlog Refinement as a preemptive measure to clarify tasks and preempt potential obstacles. The goal is to dissect each item, clarify doubts, and achieve a shared understanding, ensuring everyone is on the same page before the sprint commences.

Soon, we'll discuss the "Definition of Ready" – a set of criteria that a backlog item must meet before it's deemed fit for execution. Think of Backlog Refinement as the process to meet this definition, ensuring each task is adequately understood and primed for successful execution.

SPRINT REVIEW

Diving into the Sprint Review, it serves as a retrospective showcase of the team's endeavors over the sprint duration, typically two weeks. As we previously mentioned, think of it as an interactive demo session, an opportunity to present what has been achieved, while also providing a glimpse into potential future goals.

The Review fosters transparency, allowing stakeholders to see firsthand the results of the team's efforts. This helps in bridging gaps between expectations and deliverables and in bolstering stakeholder trust. While the primary aim is to demonstrate completed tasks, it's also a platform to solicit stakeholder feedback and potentially gather new insights or ideas. This doesn't necessarily mean that every piece of feedback will culminate in an action item, but it's crucial to be receptive and open to suggestions.

There will be instances when the accomplishments of a sprint aren't overtly visible. For example, backend infrastructure work or design blueprints might not present as tangible products. In such scenarios, the team should endeavor to articulate and represent their work in a way that's comprehensible to stakeholders, emphasizing the value and progress made.

The essence of the Sprint Review is two-fold: validation and course correction. Continual validation ensures alignment with stakeholder expectations and identifies potential deviations early on. Small, frequent adjustments are more manageable and less disruptive than significant shifts after prolonged periods. Imagine navigating a journey – it's more efficient to adjust your course continually based on real-time feedback than to discover you've been off-course only after reaching your destination. Some may liken the Sprint Review to a Design Review.

While there are similarities, the latter can be more formal and rigorous. The emphasis for Sprint Reviews should be on agility and collaboration. While reviews are indispensable, it's essential to streamline them, ensuring they're efficient and not cumbersome. Engaging stakeholders earlier in the design and development process can preempt the need for extensive rework post-review. After all, the objective is to get it right from the onset, reducing the necessity for extensive revisions later. The proactive approach trumps the reactive, and that's the ethos behind the Sprint Review.

RETROSPECTIVE

The Retrospective stands out as Scrum's most pivotal event. While its significance is unmatched, simplicity is its essence. For a two-week sprint, an hour-long Retrospective typically suffices. Anything less may not only feel rushed but might compromise the value derived from it.

The Retrospective's core revolves around three primary queries: What succeeded? What faltered? How can we enhance? While these questions are foundational, their presentation can vary. Numerous engaging techniques, like the Starfish or Sailboat methods, infuse creativity and fun into the Retrospective. Shifting methodologies and game-like approaches not only reinvigorate the team but also promote varied perspectives.

Occasionally, forgoing the traditional structure for a more relaxed interaction, like a virtual coffee session or happy hour, can be beneficial. It's a reminder that the objective isn't solely product or process enhancement. Equally pivotal is fortifying the team's rapport and individual growth. Such casual interactions promote camaraderie, trust-building, and deeper understanding among team members. These seemingly minuscule investments in team health yield exponential dividends, leading to a harmonious team that produces superior results.

While it's natural to emerge from a Retrospective with a plethora of improvement ideas, moderation is key. Prioritizing one to two actions, three at most, ensures the team doesn't spread itself thin over a two-week span. Instead of attempting broad, sweeping changes, focusing on smaller, tangible improvements can substantially amplify team productivity. It's about making consistent, incremental progress, ensuring that each step taken is deliberate and beneficial.

SCRUM EXPRESS QUIZ QUESTIONS

1. **A list of ALL the requirements the customer wants (even those not worked on) is known as ________**
   A. Product backlog
   B. Sprint backlog
   C. Scrum
   D. Sprint

2. **A list of SOME of the long list requirements (only those to be done in a sprint) is known as ________**
   A. Product backlog
   B. Sprint backlog
   C. Scrum
   D. Sprint

3. **The product backlog is prioritized and owned by __________**
   A. Product owner
   B. Developers
   C. Scrum master
   D. Scrum team

4. **The first event of a sprint is ________**
   A. Sprint planning
   B. Backlog refinement
   C. Daily scrum
   D. Retrospective

5. **The average length of a sprint (used in most firms) is _________**
   A. 1 day
   B. 2 days
   C. 2 weeks
   D. 1 month

6. The end-result of the sprint is known as ________

A. PSI
B. PMI
C. DCI
D. PSO

7. A daily-scrum is __________

A. A planning meeting
B. A status meeting
C. A phase gate
D. Resolution meeting

8. Priming the backlog and preparing it for the NEXT sprint through discussions and re-prioritization actions is known as ______________

A. Backlog planning
B. Backlog refinement
C. Backlog vision
D. Story gliding

9. The potentially shippable increment should be shown to the customer in a demonstration. This is known as __________

A. Retrospective
B. Scrum Zero
C. Sprint Review
D. Backlog refinement

10. The Scrum Team should spend time in self-reflection and checking itself for optimization and improvements. This is known as ________

A. Retrospective
B. Sprint review
C. Lessons learned meeting
D. Carrot and stick aversion

11. You are working on a team using Scrum. The timebox for your work is being decided. Which of the following is the likely timebox?

A. 2 weeks
B. 2 months
C. 4 months
D. 1 year

12. You are working on a team using Scrum. The purpose of the ________ is to inspect the outcome of the Sprint and determine future adaptations.

A. Sprint review
B. Sprint retrospective
C. Daily standup
D. Backlog refinement

13. You are in the process of a demo. Stakeholders are keen on seeing the product in operation. How long should this process take for a 1-month sprint?

A. 1 hour
B. 4 hours
C. 12 hours (1.5 day)
D. 15 minutes

14. In plan-driven project management, which of the following is typically flexible?

A. Schedule and budget
B. Schedule
C. Budget
D. Scope

15. In Agile, which of the following is flexible?

A. Schedule and budget
B. Schedule
C. Budget
D. Scope

SCRUM EXPRESS QUIZ ANSWERS

1. A list of ALL the requirements the customer wants (even those not worked on) is known as __________

A. Product backlog

2. A list of SOME of the long list requirements (only those to be done in a sprint) is known as __________

B. Sprint backlog

3. The product backlog is prioritized and owned by _____________

A. Product owner

4. The first event of a sprint is __________

A. Sprint planning

5. The average length of a sprint (used in most firms) is ___________

C. 2 weeks

6. The end-result of the sprint is known as _________

A. PSI

7. A daily-scrum is ___________

A. A planning meeting

8. Priming the backlog and preparing it for the NEXT sprint through discussions and re-prioritization actions is known as ________________

B. Backlog refinement

9. The potentially shippable increment should be shown to the customer in a demonstration. This is known as ___________

C. Sprint Review

10. The Scrum Team should spend time in self-reflection and checking itself for optimization and improvements. This is known as _________

A. Retrospective

11. You are working on a team using Scrum. The timebox for your work is being decided. Which of the following is the likely timebox?

A. 2 weeks

12. You are working on a team using Scrum. The purpose of the __________ is to inspect the outcome of the Sprint and determine future adaptations.

B. Sprint retrospective

13. You are in the process of a demo. Stakeholders are keen on seeing the product in operation. How long should this process take for a 1-month sprint?

B. 4 hours

14. In plan-driven project management, which of the following is typically flexible?

A. Schedule and budget

15. In Agile, which of the following is flexible?

D. Scope

CHAPTER 5

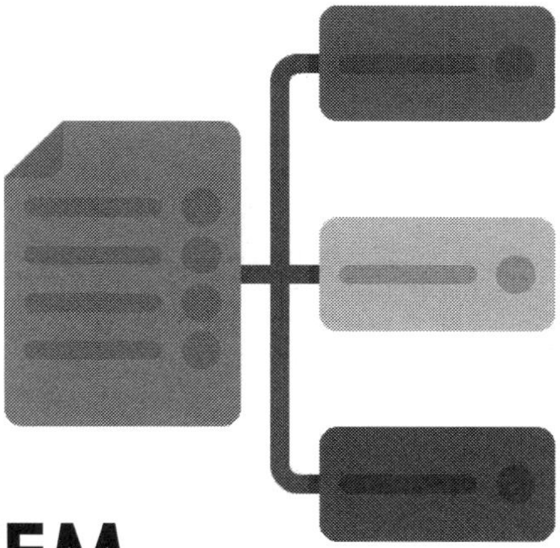

PRODUCT BACKLOG ITEM MANAGEMENT

DEFINING THE PRODUCT

In the realm of Agile, the myth that we sidestep traceability couldn't be further from the truth. Those foundational project management elements we once relied upon remain integral, albeit expressed differently. One such approach lies in delineating the hierarchy of the product under development. At its core, it's about requirement traceability, a structured breakdown of work.

Foremost, we pivot from a project-centric to a product-centric view. While projects operate within confined timeframes, products boast extensive lifecycles encompassing inception, evolution, refinement, and support phases. This temporal distinction, with projects being fleeting and products enduring, sets the stage for our hierarchical considerations.

The product tier encapsulates the grand vision and mission. It's the overarching description of our end-goal, delineated by its purpose and the rationale behind its inception. The emphasis on the "why" is paramount, for it provides clarity and connects us with the broader objective.

Subsequent to the product level, we venture into "epics." Epics, by definition, are sizable segments of functionality, more attuned to business value than mere operational steps. They lay the groundwork for value propositions rather than procedural phases, setting the stage for what follows in the hierarchy.

With a value-centric approach, Agile redefines work organization. Instead of categorizing tasks by phases or steps, Agile spotlights potential value to the customer. This focus realigns our priority matrix. Value delivery tops the list, followed by sequencing and other considerations. To encapsulate, in Agile, "epics", "features", and "stories" are not just tasks or phases. They're embodiments of potential value propositions we offer to our clientele.

The term "potentially" takes on significance when we address value. Simply putting forth an element of a larger system doesn't guarantee immediate value for end-users. Consider the scenario of an online banking portal. While an operational login screen holds inherent value, its existence in isolation doesn't translate to direct utility for the user. An amalgamation of various functionalities culminates in a release that truly serves user needs.

This foundational release in the agile world is known as the MVP or Minimum Viable Product. It represents a paradigm shift from bloated product delivery to streamlined functionality that aims for maximal impact. The MVP philosophy is straightforward: Deliver the least, yet most crucial set of features to ensure customer satisfaction. It's a nod towards achieving optimal outcomes through minimal efforts.

However, the interpretation of MVP varies between the software and tangible product domains. Software, given its malleable nature, allows for iterative additions post the initial release. Contrastingly, tangible products often demand a more comprehensive feature set in their MVP, given the complexities and costs tied to subsequent iterations.

Nevertheless, the MVP's underlying principle remains consistent across sectors: Prioritize speed and quality in product releases to ensure faster returns and satisfied customers. By focusing on the essential, we pave the way for a more responsive, adaptive, and efficient product lifecycle.

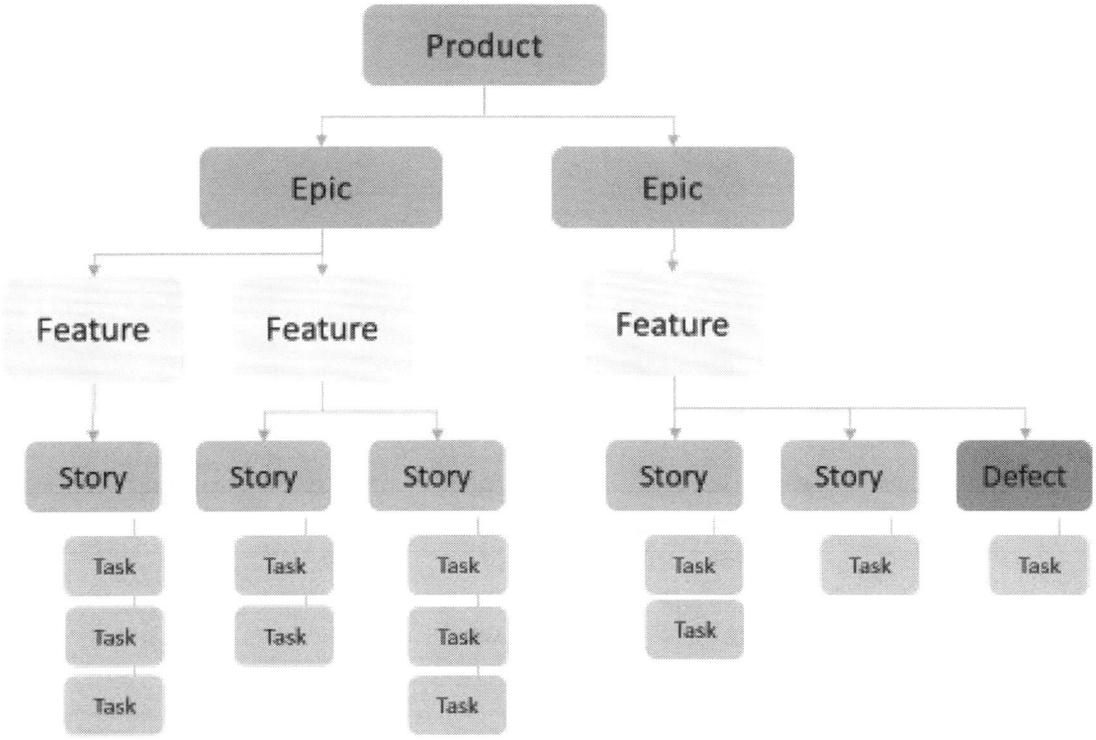

Figure 12:Product Backlog Item Breakdown

This arrangement is just one way to break the work down.

A mnemonic: **Phill – Eats – Fries – Sometimes** on **– Tuesdays**

Product – Epic – Features – Stories - Tasks

The hierarchical structure in Agile methodologies like Scrum offers a systematic approach to product development, ensuring clarity, priority, and value delivery. Here's an elaboration on each:

1. **Product**: The overarching entity. It's the end-to-end solution or system we aim to develop, refine, and maintain. It encompasses the vision, the mission, and the overall value proposition.
2. **Epic**: Situated beneath the product level, epics represent larger portions of functionality. While they're valuable, they're often too extensive to be developed within a single sprint. Epics are essentially collections of related features that together fulfill a broader user need or business goal.
3. **Features**: Subsections of epics. Features are more granulated functionalities that, while smaller than epics, might still be too comprehensive for completion in a single sprint. They outline specific functionalities or capabilities that the product should have.
4. **User Stories**: Nested under features, user stories break down features further into smaller, digestible units of work. These stories are designed to be achievable within a single sprint. They're user-centric, often articulated from the user's perspective, ensuring that what's developed aligns with user needs. Alongside user stories, you might find other units like Technical Debt or Defects – essential pieces to address but might not directly align with new value creation.
5. **Tasks or Subtasks**: The most granular level in this hierarchy. Tasks specify the nitty-gritty details of how a user story will be implemented. They are the step-by-step actions or processes necessary to bring a user story to life. While stories focus on the "what", tasks are all about the "how".

When you look at the spectrum from Product to Tasks, it becomes clear that it's a journey from a macro vision to micro execution. This hierarchy not only provides teams with a clear roadmap but also ensures that every bit of work aligns with the bigger picture and delivers real, tangible value. It's like constructing a building, starting with a blueprint (Product/Epic), then identifying rooms (Features), designing each room layout (User Stories), and finally, choosing the paint colors, fixtures, and other details (Tasks). Each level has its own purpose and importance in the broader context of product development.

TYPES OF PRODUCT BACKLOG ITEMS (PBIs)

In Agile development, particularly Scrum, the Product Backlog is the single authoritative source for things you'll work on. These items on the backlog are called Product Backlog Items (PBIs). Each PBI represents a piece of functionality or value to be added to the product. There are a few types of PBIs, and they each serve different purposes:

1. **User Story**:
    - This is the most commonly encountered PBI.
    - A User Story centers on end user value, capturing requirements from the user's perspective. The canonical format is: "As a [type of user], I want [some feature] so that [some benefit/a value]."
    - While stories aim to be small enough to be completed within a single sprint, they should always be framed in a way that focuses on delivering user value.
2. **Spikes (or Enablers)**:
    - Spikes are research-oriented PBIs.
    - Their primary goal is to gain knowledge, address uncertainties, or eliminate potential risks.

- While they don't produce direct, tangible value like a User Story might, they are vital for guiding the team's future actions or for the preparation of future sprints.
- A Spike's outcome is often insights, data, or clearer understanding which can lead to the creation of new PBIs.

While User Stories focus on "what" to build, Spikes tend to answer the "how" or "if" questions. Both are critical components of a well-rounded backlog.

In summary, the Product Backlog is a dynamic tool, constantly evolving and adapting to the product's needs. As teams progress, new learnings from Spikes, feedback on delivered User Stories, and changing business conditions all contribute to its continuous refinement. This flexibility and adaptability are the hallmarks of Agile, ensuring that teams are always focused on delivering the highest value in the most effective way.

TECHNICAL DEBT AND PRIORITIZATION

Technical Debt, often described as the "cost" of postponed changes in a software system, is an inevitable aspect of the development process. Over time, as teams build products, certain decisions made—whether due to time constraints, resource limitations, or changing requirements—can result in accumulated technical debt.

1. **Nature of Technical Debt**:
   - It isn't inherently "bad." At times, accruing some technical debt might be a strategic decision to meet a deadline or a particular business goal.

- However, if not addressed, it can lead to increased maintenance costs, reduced agility, and potentially more significant problems down the road.

2. **Balancing Priorities**:
    - Prioritizing features over addressing technical debt is a common challenge. While product areas often crave new features, ignoring technical debt can slow down future feature development.
    - Discussing this balance is essential. Stakeholders need to understand that by allocating time to reduce technical debt, the team can work more efficiently and innovatively in the future.
3. **Defects**:
    - Defects are inevitable in any product development. Like technical debt, they need prioritization. Critical defects that affect usability or functionality typically take precedence.
    - Some organizations might have dedicated teams addressing defects, while others include it as part of the regular backlog, managed by the same team handling new features.
4. **Team Organization and Strategy**:
    - How a team is structured and organized can influence their efficiency in managing technical debt and defects.
    - Depending on the product's needs, having a singular team address both new features and maintenance might be beneficial. In other situations, segregating these responsibilities across teams may be more efficient.

In essence, effective product management is about understanding trade-offs. While the allure of new features is undeniable, acknowledging and addressing technical debt and defects is crucial for long-term success. Regular discussions and prioritization exercises can ensure teams strike the right balance, driving product growth while maintaining its foundational integrity.

PRODUCT BACKLOG ITEMS AT STORY LEVEL

Let's begin with a recap of the product definition and the product hierarchy. When In Agile methodologies, requirement traceability is pivotal. This ensures we adhere to both good engineering and project management practices. Here's an overview of the product hierarchy:

1. **Product Focus**:
   - A product is what you're building and continually enhancing, not to be confused with projects, which may involve different facets of a product.
2. **Epics**:
   - These are large chunks of work, usually too vast to be tackled within a sprint. By fragmenting the product, it facilitates prioritization and handling.
3. **Features**:
   - These are more manageable pieces, ideally suited for sprints. The key is to delineate them so their progress can be easily monitored.
4. **Stories**:
   - These are concise units of work designed for sprints. Ensuring they fit within a sprint's timeframe is crucial for effective planning and execution.
5. **Tasks/Subtasks**:
   - These are detailed steps required to execute stories. Though primarily for the team's reference, they help streamline work processes.

VARIETIES OF PRODUCT BACKLOG ITEMS (PBIs)

1. **User Stories**:
   - These cater to end users and stakeholders, providing features that offer direct value.

2. **Spikes/Enablers**:
    - These are meant for research or prototyping, pivotal for knowledge acquisition and risk mitigation.
3. **Technical Debt**:
    - A result of shortcuts or workarounds taken in the development process. It's crucial to allocate time to address these to maintain product quality.
4. **Defects/Bugs**:
    - These are functionalities that aren't working as intended. While Technical Debt might be operational but not standard, a defect is outright non-functional.

By understanding and organizing work using this hierarchy, Agile teams can manage tasks effectively, ensure product quality, and deliver value consistently.

WHAT IS A USER STORY?

User Stories are very useful. The idea with the User Story is that it's basically a description of a requirement, something that's needed. As the name would imply, the idea is that it's a Story about a user's need. We like telling stories. Stories identify a problem. It helps you connect to your user to your customer. Therefore, Stories are really beneficial. It helps you feel the problem and understand the problem better when you're telling a Story. So instead of just saying it's a system shall kind of requirement, try to tell who it's for, try to state the problem, and give better context. The better context is going to help the team to build the thing. The system shall requirements; they're okay. We've done this for a really long time, but they don't give you the full context.

A User Story is a brief description of a need or problem:

User stories are short, simple descriptions of a feature told from the perspective of the person who desires the new capability, usually a user or customer of the system.

Now, there are three aspects of a User Story. My User Story has the aspects of the card, the conversation, and the confirmation. Those are the three C's. The card is the initial User Story; it's written description.

3 aspects of user stories:

- *Card*: A written description (often on an index card)
- *Conversation*: Conversations about the story to clarify it and flesh out the details
- *Confirmation*: Criteria that can be used to determine when the story is completed

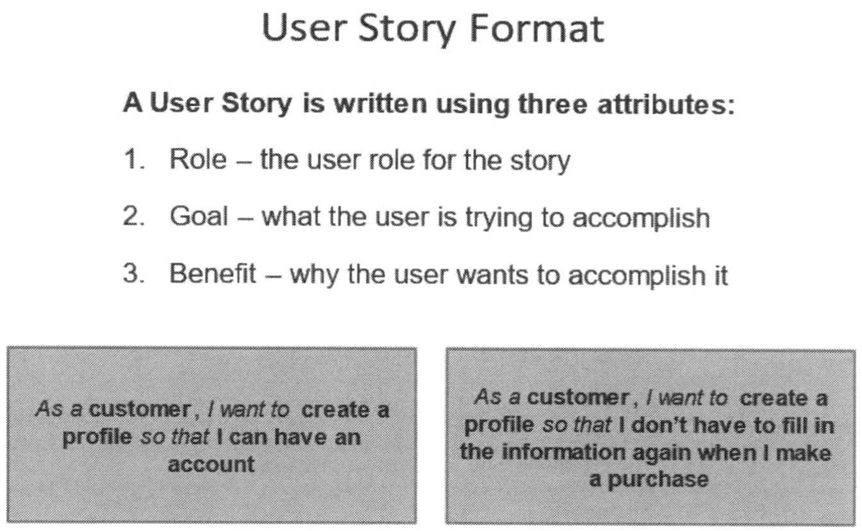

FIGURE 13: USER STORY FORMAT

We actually often do put these on index cards or **Post-it®** notes. Obviously, it's not necessary to do that, especially when you're remote, not very useful. But the idea is you write them on a very small ***Post-it®*** note or index card. The reason for that is that we don't want you to list all of the details upfront.

This is an insightful and well-rounded description of the user story process in an Agile development framework. It clearly highlights the three Cs of User Stories: Card, Conversation, and Confirmation.

1. **Card**:
    - This is the initial step where a basic idea is jotted down. It's like planting a seed. It's not about precision but about capturing an idea for a need.
    - It represents a high-level requirement or a feature that provides value to the end-users.
    - The card is more of a placeholder for a deeper conversation that will happen later.
2. **Conversation**:
    - It's in this phase where the team dives deep into the details of the requirement. The aim here is to understand the problem and explore possible solutions.
    - The Product Owner plays a pivotal role in this conversation, providing context, vision, and clarity on the requirement.
    - The development team can propose alternative solutions and improvements, ensuring that they're building the right thing in the right way.
    - Quality starts here. It's essential to understand and refine requirements as this can significantly affect the project's success.

3. **Confirmation**:
    - This is where the acceptance criteria come into play. These criteria are the measures by which the implementation of the user story will be judged.
    - They ensure that the team and the Product Owner are on the same page regarding the expected outcome.
    - They serve as a clear boundary, preventing scope creep and helping maintain a focused approach to implementation.

In essence, User Stories are not just requirements but are tools for collaboration, ensuring that the team is aligned with the stakeholders' needs and values. They emphasize the importance of continuous communication and feedback, promoting a shared understanding of goals. Remember, Agile is not about doing things faster but about delivering the highest value and ensuring that the team is building the right thing.

USER STORY FORMAT

The traditional User Story format provides a structured way to capture requirements, focusing on the role, the goal, and the benefit. This approach can be employed beyond just Agile projects. Even organizations that use traditional models can benefit from this format, leveraging the concept of progressive elaboration which is central to Agile

The User Story usually goes like this: "As a [type of user], I want [an action] so that [a benefit/a value]". This is not just about specifying a requirement. It underlines the user's purpose and what they aim to achieve, thereby spotlighting the value. One can argue that the 'why' – the value or the benefit – is the most crucial aspect. If the value of a requirement isn't clear, perhaps it doesn't deserve a spot in the backlog. Conversely, being able to articulate the value can be instrumental in prioritizing tasks.

Take, for example, a software context: "As a customer, I want to create a profile so that I can have an account." This User Story appears rather vague. While it gives a general sense of the requirement, it doesn't quite clarify why a user might need an account. Contrast this with: "I want to create a profile so I don't have to fill in the information again when I make a purchase." This latter version is more specific, presenting a clear problem the user aims to sidestep.

While the User Story format provides a helpful structure, it's important to remember the goal is understanding the requirement. In some scenarios, it might be more effective to forgo the format and simply narrate the problem or scenario. Narrating a requirement as a story has its own merit. Stories can evoke emotions, making them memorable and easier to grasp. There's a technique called specification by example, wherein requirements are presented as specific problem-based examples. Such stories can be much more engaging and illuminating.

It's imperative to remember, however, that not every requirement will neatly fit the User Story mold. It's unwise to be too stringent about it. Some organizations erroneously use adherence to the User Story format as a performance metric for their teams. But just using the format doesn't guarantee the content is valuable. The essence is in the content, not the container.

There might be scenarios where the format isn't apt, particularly with very technical or internal requirements. That's perfectly alright. While the User Story format can be beneficial, the real aim should be clear understanding and communication of requirements. If the format helps, great! If not, it's okay to shift to a more narrative or different style. The objective is to get the requirements right, irrespective of the format.

ACCEPTANCE CRITERIA

The Acceptance criteria provides clarity and precision to the User Story. If you think of the User Story as setting the stage by identifying the problem and the value of addressing it, then the Acceptance criteria drills down into the specifics of how that story reaches a satisfying conclusion. It's a delineation of the conditions that must be met for a feature or function to be deemed complete. Without these criteria, the team is left to make assumptions about what the Product Owner expects, and that's where the risk of deviations and missed expectations arises.

When crafted meticulously, Acceptance criteria:

1. **Clarifies Expectations**: It ensures everyone on the team has a shared understanding of the feature's purpose and functionality.
2. **Guides Development**: It directs the developers on what exactly needs to be built and tested.
3. **Facilitates Testing**: Testers use these criteria to determine what to test for and ascertain that the feature works as intended.

For instance, if we refer back to the previous User Story: "I want to create a profile so I don't have to fill in the information again when I make a purchase," the Acceptance criteria could include:

- A user can input personal information into the profile fields.
- A user can save the profile.
- Upon return, a user's information is auto-populated into the purchase fields.
- A user can edit the profile information.

This not only specifies what needs to be built but also what needs to be tested.
A common misconception is that Acceptance criteria must contain every minute detail from the outset. While it should be comprehensive, striving for perfection from the start can lead to paralysis. Instead, the goal should be clarity and cohesion. As teams begin to work on the User Story, they'll often find that they need to adjust or add to the Acceptance criteria based on what they learn. This isn't a flaw in the process; it's a feature. Agile emphasizes adaptability and learning from doing.

That said, it's critical not to mistake vagueness for adaptability. If Acceptance criteria are too ambiguous, teams can end up building features that don't quite align with the Product Owner's vision or the users' needs. A careful balance is required: the criteria should be clear and concise enough to guide development and testing but also flexible enough to accommodate unforeseen insights or challenges that arise during the development process.

While the User Story provides a high-level overview of what is needed and why, the Acceptance criteria fills in the details of how it should work and what it should accomplish. This tandem ensures that development efforts are aligned with business goals and user needs, and it maximizes the chances of project success.

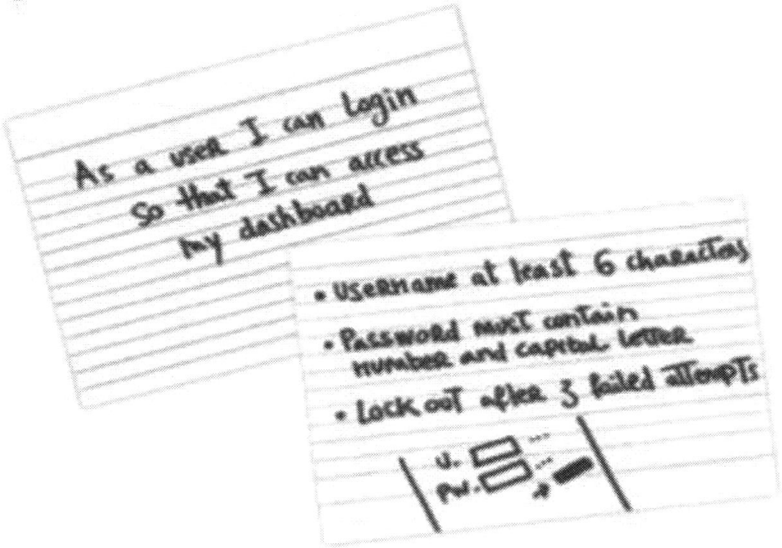

FIGURE 14: ACCEPTANCE CRITERIA CONCEPT

Embracing uncertainty and being adaptive in your approach is a hallmark of Agile methodologies. While thoroughness is encouraged, striving for perfection can lead to paralysis. Too often, teams can become bogged down in the analysis phase, hoping to address every possible scenario in their User Stories and Acceptance criteria. However, as the saying goes, "Perfect is the enemy of good."

This doesn't mean that you should approach your work haphazardly. Instead, you should focus on building a foundation, an MVP (Minimum Viable Product) if you will. This MVP should be guided by your User Story and its Acceptance criteria. These criteria provide a roadmap of essential features that your product must have to be considered complete.

Starting with a simple checklist for Acceptance criteria can be incredibly helpful. It offers a clear and straightforward way to communicate the requirements. By using phrases like "I know that we're done when," you're offering a user-centric perspective, ensuring that the outcome meets the user's needs.

But for those who crave a bit more structure or wish to dive into the nitty-gritty details, the Gherkin format offers a more in-depth approach. The "Given-When-Then" format of Gherkin provides a structured yet intuitive way to describe a feature's behavior without detailing how that functionality is implemented. This makes it particularly well-suited for behavior-driven development (BDD).

While not every situation may be ideal for Gherkin, its structured nature and emphasis on conditions and outcomes can make it easier to turn Acceptance criteria into testable specifications. For teams that utilize automated testing, this can be a boon, streamlining the transition from criteria to test cases.

As for documentation, while thorough documentation is crucial, it needn't be ornate or cumbersome. Often, simple low-fidelity mock-ups can convey ideas effectively. Sketches, whiteboard drawings, or even paper prototypes can serve as valuable tools to visualize and iterate upon design concepts.

In summary, while thorough preparation and documentation are essential, they shouldn't impede progress. Whether you choose a simple checklist or the Gherkin format for your Acceptance criteria, what's crucial is that it clearly conveys the desired outcome and sets the team up for success. Strive for clarity, remain adaptive, and remember that the end goal is a product that meets users' needs and delivers value.

PRODUCT BACKLOG

The Product Backlog is essentially a prioritized "to-do" list. However, unlike a regular to-do list, items (or requirements) in this backlog can evolve or become obsolete. It's not unusual for some requirements to sit in the backlog for months, and in that time, the overall vision or scope of the project might change. If a requirement no longer aligns with the current project vision, it should be reconsidered or even removed.

The responsibility of maintaining this backlog lies with the Product Owner. They determine what's essential, what's not, and in what order tasks should be tackled. But remember, while the Product Owner has final authority, anyone can suggest items. In fact, the entire team is encouraged to contribute since they often have valuable insights and ideas. However, the ultimate decision of what makes the cut remains with the Product Owner.

Adding new items or reprioritizing the backlog might sound like scope creep, which can lead to project risk. To manage this, adopt the principle of "bring something in, move something out." In other words, if a new requirement is added, another might need to be de-prioritized or even removed to ensure the scope remains manageable.

SPRINT BACKLOG

The Sprint Backlog is a subset of the Product Backlog, representing the tasks chosen for the upcoming sprint, typically lasting around two weeks. The selection is a collaborative effort between the Product Owner and the team. The aim is to determine a realistic and achievable set of tasks that align with the sprint's goal.

Overcommitting is a trap many teams fall into, leading to potential quality issues and burnout. The concept of velocity, which measures how much a team can handle consistently over sprints, is a good guard against this. Stick to this metric, and remember that it's always better to under-commit and then pull in more work if the team finishes early. Conversely, constant changes to the Sprint Backlog can derail the team's focus, so it's crucial to keep it stable throughout the sprint duration.

In the event of unforeseen essential tasks (like a production emergency), the team might need to adapt, but this should be an exception, not the norm. The Scrum Master plays a crucial role here, shielding the team from unnecessary disruptions and ensuring they can concentrate on their designated tasks.

QUESTION 20

Which of these BEST describes a product backlog?

A. Chunk of work to be performed by the team

B. Prioritized list of all work to be performed by the team

C. Prioritized list of all tasks to be performed by the team in a month

D. A backlog of all work completed by the team in the last sprint

That could be considered an Epic or Feature. It's a collection of work to be done by the team, and this could also be a little bit confusing. It could also potentially mean the sprint but, in this question, it's more likely the Epic or the Feature, the chunk of work that we're going to execute.

QUESTION 21

Which of these describes an EPIC?

A. A big chunk of functionality or work
B. A phase of tasks
C. A feature or task
D. A user story

The answer is A. A big chunk of functionality or work. I can understand why D might be a little bit confusing, but consider this, an Epic is really a collection of all of those User Stories. Hence, it's more of a container that describes a high-level chunk of functionality that we want to try to produce. While the User Story is a very, very small increment of functionality that can comfortably fit inside of a sprint. The real difference is the size of the work. You could potentially consider that an Epic is really just a very, very, very large User Story. It's the same thing, a chunk of functionality but a User Story...it's specifically User Story because it's small and it fits in the sprint.

Product is always at the top because that's the thing that you're ultimately building, a product or service. And then below that, there are Epics and Features and then finally Stories and other types of things similar to Stories.

QUESTION 22

Which of the following is the correct hierarchy?

A. Epic – Product –Feature – Story - Task
B. Feature – Product – Epic –Story – Task
C. Product – Epic – Feature – Story – Task
D. Story - Product – Epic – Feature –Task

Remember, the Product is always at the top of the hierarchy. That's the main way that you can tell this. All of these elements are correct in every one of these responses or selections, but the important thing is the Product comes first. Epics are big chunks, Features are somewhat smaller, and Stories are very small. Then, the Tasks, of course, are the steps to get there.

As a blank, I want blank, so that blank is the role, the goal, and the benefit. You are really helping everybody out here,

Answer: Product – Epic – Feature – Story – Task

QUESTION 23

Which of these describes a user story format

A. want A, so that B, as a C
B. As a X, I want Y, so that Z
C. So I can A, as a B, I want C
D. So that X, As a Y, I want Z

The correct answer is "as a blank, I want blank, so that blank." But I would like to say this as well; we don't want to be dogmatic about the format. I do have some teams that decide to change it up because they view the most important thing is the "so that" statement, so they like to put it first. That way, it's always the first thing they put down; the problem statement is number 1. That's the most important thing to them, so they just change the format around to remind them. What tends to happen to some behaviors I've seen with the User Story format is, if you stick with the "as an I want so that." What usually winds up happening is people forget or just ignore the "so that." They just don't add it to their User stories. They just get the first two parts. Doing that is hard. It's really not an easy thing to do. That's the important stuff, and it's very difficult to state this, so people often ignore it. It may be beneficial to you to change the format up and maybe put the "so that" statement first. That's okay. You can do that.

Answer: As a X, I want Y, so that Z

QUESTION 24

Which of these describes technical debt?

A. Something that benefits the user directly, or enables a future benefit

B. Research, prototype, risk reduction, knowledge acquisition

C. Things we should have done correctly the first time, but likely took a shortcut to meet a deadline

D. Something that is causing an impact and needs to be fixed

Think about the word debt. Debt is where you're basically borrowing from now, and then you're going to have to pay it back in the future. And that borrowing is doing a shortcut to meet a deadline, but you've got to pay it back and fix it eventually. Technical Debt is things that we should have done the first time correctly, but perhaps we had to take a shortcut; it might have been intentional or unintentional. It's basically stuff that's working but needs to be fixed eventually. Otherwise, it could potentially cause harm to our product. The answer is C.

Answer: Things we should have done correctly the first time, but likely took a shortcut to meet a deadline

QUESTION 25

Agile has built-in mitigation techniques:

A. Schedule, WBS, Risk Register
B. Sprint Review, Prioritized backlog, Daily Scrum
C. Frequent Releases, WBS, Phase Gates, Toll Gates
D. Kill Points, Kano, Transparency, Decentralized decision making

We don't do Toll Gates, we don't do Phase Gates. We don't technically do WBS, but we do things similar to this. We don't actually have them defined in the process, so that's why A would not be the right answer. We pay a little less attention to things like WBS, which we don't do. We don't technically have A Risk Register either, Although there is going to be some discussion about managing risk, and there are some things that we can make use of similar to that. We'll talk about how to manage that.

Answer: B. Sprint Review, Prioritized backlog, Daily Scrum

PRODUCT INCREMENT

The Product Increment refers to the tangible output or result produced at the end of a sprint. The essence of this concept is to ensure that after each sprint, there's a functioning piece of the product. It's an embodiment of the agile principle that values a working product as the primary measure of progress.

Key Points:

1. **Nature of Product Increment**:
    - It is the part of the product that's been developed during the sprint.
    - Ideally, it should be in a condition that's releasable to customers, but it doesn't always have to be.
2. **Shippable Doesn't Always Mean Shipped**:
    - While the aim is to produce something shippable at the end of a sprint, it doesn't mean it's always shipped to the customers. Especially in cases involving tangible products, multiple sprints might be required to create a shippable increment.
3. **Done Means Done**:
    - In Agile, an item is either complete or not. There's no in-between. The focus is on delivering working products, not just completing tasks or phases.
4. **Software Releases**:
    - The software domain allows more frequent releases, but there's a cost associated with each release.
    - Internally, there are processes and validation steps that may be resource-intensive.
    - Externally, frequent updates can inconvenience or even annoy users, potentially alienating them.
5. **Balancing Releases**:
    - While agile encourages faster release cycles, it's crucial to find a balance.
    - An overzealous approach to releases, as seen with products like Adobe Acrobat Reader in the past, can frustrate users.

While the agility of product development and release is beneficial, it's equally vital to assess the impact on both internal processes and external users. Proper pacing and balance are essential to avoid overwhelming or annoying customers.

DEFINITION OF DONE

Definition of Done (DoD) is a shared understanding among the team members and the Product Owner of when a task or product increment is considered completed. It's a set of criteria that a unit of work must fulfill for it to be marked as done.

Key Points:

1. **Purpose of DoD**:
    - It serves as a contract or checklist ensuring that quality and good engineering practices are maintained.
    - It clearly outlines the expectations, ensuring that all team members are on the same page.
2. **Content of DoD**:
    - The DoD may include various criteria like code reviews, documentation updates, user acceptance testing, or any other standard the team sees fit.
3. **Benefits of DoD**:
    - Promotes consistent and high-quality work.
    - Ensures a shared understanding and expectation across the team.
    - Eliminates ambiguity, ensuring that everyone knows the criteria that need to be fulfilled.
    - Helps prevent the "done BUT" scenario where work is claimed to be done except for one or more tasks.
4. **Difference Between User Story and Task**:
    - A User Story is a feature or functionality the team is aiming to build.

- A Task represents steps or actions to achieve that user story.
- A User Story isn't considered done unless all associated tasks are complete.

5. **Documentation & Storage**:
    - It's crucial to document the DoD and make it accessible to all team members.
    - Team spaces, like Confluence (an Atlassian product integrated with Jira), can be an excellent place to store such team agreements.

The Definition of Done is a pivotal tool in the agile methodology, ensuring that work is completed to a consistent standard. It removes ambiguity, ensuring everyone knows when a task is genuinely complete, and promotes teamwork and quality in deliverables.

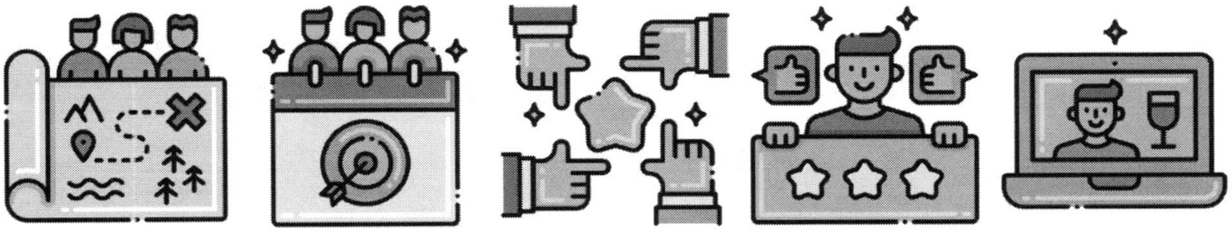

DEFINITION OF READY

Definition of Ready (DoR) stands in contrast to the Definition of Done. While DoD stipulates when a task or product increment is completed, DoR sets the criteria for when a task or product increment can begin.

Key Points

1. **Purpose of DoR**:
    - It serves as a precursor agreement ensuring that any task or product increment has reached a certain level of clarity before the team starts working on it.

- It's essentially a quality gate that protects the team from beginning work on ill-defined or premature tasks.

2. **Content of DoR**:
    - Common criteria in the DoR might include: clear understanding of the task, explicit acceptance criteria, accurate task estimation, and clarity on any dependencies.
3. **Benefits of DoR**:
    - Ensures a shared understanding of a task before it begins.
    - Reduces the risk of ambiguity and rework by making sure tasks are adequately refined before being taken up.
    - Strengthens the collaboration between the team and the Product Owner, reminding both of the importance of preparation.
4. **Emergencies & Exceptions**:
    - Reality check: Not every situation will fit neatly into these criteria. Emergencies, for instance, might bypass the DoR.
    - In case of valid emergencies, the team should act quickly without being bogged down by processes. However, such exceptions should remain rare.
5. **Relation to Traditional Methods**:
    - DoR and DoD can be seen as lightweight Toll Gates, ensuring quality at both the beginning and the end of a task.
    - Agile doesn't discard all practices from traditional or waterfall methods. Instead, it streamlines them, making them more adaptable and efficient.

The Definition of Ready is crucial in agile methodologies, ensuring that tasks are adequately defined and prepared before the team starts work on them. It helps maintain a steady workflow and minimizes disruptions due to ill-defined tasks. Both DoR and DoD, when used effectively, can greatly enhance the efficiency and quality of an agile team's outputs.

CHAPTER 6

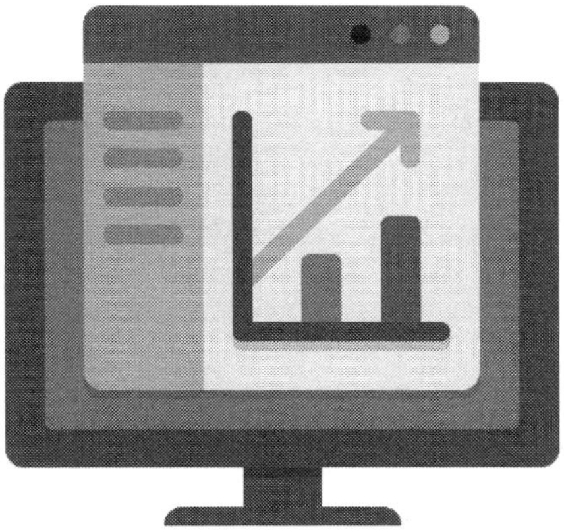

CATEGORIZATION & PRIORITIZATION

WHAT IS VALUE DELIVERY?

Value Delivery is central to agile methodologies and focuses on maximizing the benefits for the end-users or customers by concentrating on what matters most. Delivering the highest value features or functionalities first ensures that even if a project gets disrupted or stopped early, the most critical parts have already been addressed.

Key Takeaways:

1. **Shift in Focus**: Traditional project management methods may prioritize tasks based on urgency or criticality. In contrast, agile focuses on direct value to the end-user.
2. **Value Over Features**: In agile, it's not just about completing a set list of features but rather ensuring that each feature or task directly contributes value to the user. If something doesn't bring tangible benefits, its importance is questioned.

3. **Early Delivery of Value**: By focusing on the most valuable tasks first, projects can ensure that they deliver the most important functionalities early. This approach provides a significant advantage, especially in projects where timelines might be tight or resources limited.
4. **Flexibility**: By prioritizing value, teams can make informed decisions about what to include and what can be left out if time or budget constraints arise. This flexibility ensures that even with cuts, the most essential and valuable components of a project are preserved.
5. **Decision-making Guided by Value**: With a clear understanding of what provides the most value, teams can make more informed decisions throughout the project lifecycle. This focus can influence everything from feature development to testing priorities.

Value Delivery emphasizes the need to prioritize tasks and features based on their real-world impact and benefits to the end-users. It shifts the project's focus from just ticking off a to-do list to ensuring that every action taken has a purpose and adds value. This approach not only increases efficiency but also enhances customer satisfaction, ensuring that they get what matters most to them at the earliest.

DELIVER VALUE EARLY – EAT TREATS FIRST

The Concept: In life, many of us were taught to save the best for last. "Eat your vegetables first, then you can have dessert." But in the world of agile product development, the mantra is reversed. It's all about delivering the treats – the most valuable and delightful parts of a project – as soon as possible.

Why it Matters:

1. **Customer Satisfaction**: By delivering the features customers are most excited about first, you immediately provide them with tangible benefits. This can boost customer morale and confidence in the product or service you're developing.
2. **Risk Mitigation**: Tackling the most valuable features first reduces the risk of them not getting done should the project face unforeseen challenges down the road. It also allows for early feedback, which can inform and improve subsequent development.
3. **Financial Benefits**: For businesses, releasing high-value features can lead to early monetization opportunities. If customers see value right away, they're more likely to invest or continue investing in a product.
4. **Motivation and Morale**: For development teams, seeing the impact of their work early on can be a significant morale booster. It provides a clear line of sight between their efforts and the value being delivered.
5. **Adaptability**: By focusing on high-value features early on, there's greater flexibility later in the project. If changes or pivots are required, the essential value has already been delivered.

Nice-to-Have vs. Must-Have: It's essential to strike a balance. While the must-have features are non-negotiable and crucial, occasionally sprinkling in some nice-to-have features can delight customers and differentiate a product in the market. The key is to ensure that the overall product vision remains clear and that the core value proposition isn't diluted.

Scope Flexibility: In agile development, scope is considered variable. While we start with an initial scope in mind, the iterative nature of agile allows for adjustments based on feedback, changes in the market, or other unforeseen factors. This flexibility ensures that the end product remains relevant and valuable. But remember, while scope may be variable, value delivery remains constant. Always prioritize by value.

"Eating your treats first" in product development is all about ensuring that the most valuable, impactful, and often delightful features are delivered to customers as early as possible. It's a paradigm shift that ensures maximum value delivery, reduces risks, and fosters an environment of continuous feedback and improvement. So, in the world of agile, it's okay to indulge in dessert first, as long as it brings the most value to the table.

ORDERING THE BACKLOG

In the dynamic landscape of Agile product development, prioritization plays a pivotal role. The backlog is the heart of any Agile project, and its orderly arrangement is essential to ensure seamless workflow and timely delivery.

Stakeholders and Prioritization Challenges: Stakeholders are integral to any project, bringing in varied perspectives and interests. While their importance is undeniable, catering to every stakeholder's request concurrently can lead to chaos. The lack of a prioritization mechanism can overwhelm teams and dilute the project's focus.

Organizational System Thinking: It's imperative to view an organization as a cohesive system rather than fragmented silos. For teams juggling multiple stakeholders, streamlined focus on high-priority tasks can pave the way for better efficiency and value delivery. Teams should not be subjected to the stresses of parallel 'number one' priorities.

Value-First Approach:

- **Subjectivity of Value**: Value is multifaceted and can differ based on stakeholder perspectives. While for some, it might be revenue-driven, for others, it could be about user satisfaction, regulatory compliance, or brand reputation.
- **Indirect Impact**: Value doesn't always translate directly into revenue. Enhancing user experience or improving service quality can indirectly boost organizational revenue by elevating user trust and loyalty.

Prioritization Methodologies:

- **MoSCoW**: This method breaks down items into four categories – Must have, Should have, Could have, and Won't have. This clear demarcation helps teams understand the criticality and importance of each item.
- **Kano Model**: This technique evaluates features based on customer satisfaction. It distinguishes between basic needs, performance needs, and delighters.
- **Risk-Adjusted Backlog**: This technique involves adding a risk factor to backlog items. High-risk items might be tackled earlier to mitigate potential roadblocks.
- **Cost of Delay**: This assesses the financial impact of delaying a feature or project, aiding in informed decision-making.
- **The Fire Alarm Anti-pattern**: An all-too-common scenario in many organizations is the abrupt 'emergency' flagged by a higher-up, forcing teams to divert their focus. While real emergencies are inevitable, false alarms can disrupt

workflow and undermine the structured process. Such behaviors, often stemming from an authoritative mindset, can jeopardize the team's morale and the project's overall success.

Prioritization isn't just about ordering tasks; it's about aligning teams, appeasing stakeholders, and delivering value efficiently. Educating stakeholders, especially those with significant influence, about the repercussions of circumventing the process is crucial. A well-ordered backlog, driven by value and backed by robust prioritization methodologies, ensures that teams remain on course, stakeholders stay informed, and projects achieve their intended goals.

CATEGORIZATION

Let's talk about two categorization approaches MoScoW and Kano. I actually call this categorization and not prioritization, and the reason is, that we might identify a hundred things in a project that are must-have items. These are the things that you have to deliver, or the product that you're producing is not functional. These must-haves, are normal. However, if you define a hundred of them, your job isn't finished yet because basically what we're saying, is that we have a hundred priority ones, and that's not fair to the team. So just because we've defined something as a must-have doesn't mean your job is done. You then have to rank within the must-haves to help you get closer to a priority 1, priority 2, priority 3.

MoSCoW:
- ❑Must Have – baseline requirements
- ❑Should Have – deliver if possible
- ❑Could Have – not critical
- ❑Won't Have (don't deliver) or Would Like (wouldn't it be cool if…)

Kano Analysis:
- ❑Must Have – baseline requirements
- ❑Satisfiers – the more the better
- ❑Delighters – at least one or two
- ❑Dissatisfiers – don't deliver these

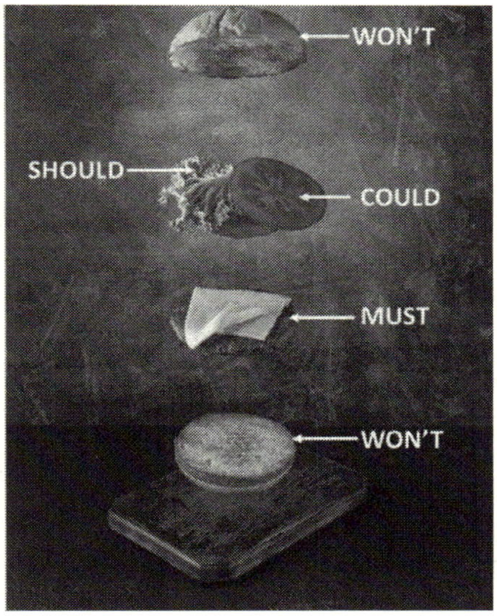

Figure 15: MOSCOW & KANO

MoScoW

MoSCoW is a technique used in project management and product development to categorize requirements and features which will ultimately ease prioritization. MoSCoW is an acronym that stands for Must have, Should have, Could have, and Won't have. The idea behind MoSCoW is to categorize requirements or features into four groups based on their priority and importance.

- Must have: These are requirements or features that are essential and critical to the success of the project or product. Without these, the project or product cannot be considered complete or successful.
- Should have: These are requirements or features that are important but not critical. They are needed to achieve the desired outcome of the project or product, but they can be delayed or omitted if necessary.
- Could have: These are requirements or features that would be nice to have but are not essential or critical. They can be considered if time and resources allow.
- Won't have: These are requirements or features that are not included in the project or product at this time. They may be considered for future releases or

versions. Or Would Like: These are innovative features that are unexpected but would delight the customer.

Let's discuss how MoScoW works. Here's an example; you're ordering a hamburger or a cheeseburger. You say that the meat in the cheeseburger is a should have, and the bun is the only must-have. That might be correct for you, but it's not correct to me. That's not what I would view as the most important thing. I'm on the Keto diet, so to me, the most important thing is going to be the meat and maybe following that could be the cheese, that might also be a must-have to me, and everything else would be a nice-to-have; maybe a could have kind of thing. And the bun, I'm not going to have that at all. That's something I don't even want because, again, carbs, right? So, consider this, my middle daughter, she's a vegan, so to her that patty; nope, doesn't want it at all. To her, the most important thing is going to be the lettuce and maybe the bun if it's a vegan bun.

What we're really saying here is that it's very, very subjective. It's going to be different for different people. What we need to try to get to at the very least is honest prioritization. Don't call something a must-have just to get it on the list. You don't need to do that. Keep in mind that the people that are prioritizing this work are your business representatives. If they feel something is important to them, they can just say "This is important to me, I'd like you to work on this," and we will. That's okay. If something's a must-have, call it a must-have. If it's a nice-to-have, a could-have, then call it that. No harm in that. If the stakeholder says, "You know, I know this is a could-have, but I really want it." No problem, we'll work on that. That's not a problem with the understanding that there's a cost, and that cost is something else of equivalent size comes out. So it's not scope creep, we're not going to allow all this stuff to pile into our release. Instead, we're going to say, "We will work on whatever you want us to work on, but consider, if you add this stuff in, some of the other stuff won't get done." There is a cost to that.

That's how we manage scope creep. Be specific about that. Maintaining that time box is really important because we can manage scope creep with that

Kano

Kano is a model used to categorize customer needs or requirements into five groups based on their impact on customer satisfaction.

- Must-have: These are expected features or functionalities. If these are missing, customers will be dissatisfied. However, even if they're present and working perfectly, they don't particularly impress the users since they are taken for granted.
- Satisfiers: These features are directly proportional to customer satisfaction. The better they function, the happier the customer. They are often compared among competitors – for instance, battery life in smartphones.
- Delighters: Features or aspects that customers weren't expecting but are pleased to discover. They can create a significant competitive advantage. However, today's delighters can become tomorrow's basic needs as markets and customer expectations evolve.
- Dissatisfiers: These are requirements that, if met, can actually lead to customer dissatisfaction. For example, if a product is too complicated to use, it can frustrate and annoy the customer.

Now, I actually prefer Kano. There's a lot more to Kano than just these categories. There's Kano analysis; there's a diagram you can go through. I typically don't go through all that. I just like the four categories that it provides. I typically don't use the category of dissatisfiers. I just say this is a bad idea and throw it out. Another interpretation of dissatisfier is if you don't deliver these things, then the customer will be dissatisfied so that you may use that as well. That's up to you if you want to use it that way, but at the very least, use the top three categories. I find them very effective.

The reason I prefer Kano is these categories kind of have an emotional feeling, kind of an emotional connection to your customer. When you implement saying this is going to satisfy our customer or this is going to delight our customer, that gets us closer to how the customer is going to feel when we implement these things. I know it may sound silly; it's not usually the stuff we do in traditional approaches. Still, we want to try to get to empathy with our customers, truly understand them, and understand what they're trying to solve. That's not necessarily easy to do, but try to build that empathy, that connection to your customers.

All right, very quickly, let's go through an example of a Kano. So, let's say that I'm buying a car. Some of the must-have features might be some way of steering it. A way of stopping is probably just as important, So those are things that are must-haves, some way of making go, a motor, an engine, or some other way of making a go. Now, also recognize the language I'm using. I'm not saying a steering wheel. I'm not saying a brake pedal. I'm not saying an engine because those are solution specific words. We don't want to solve the problem in our requirements. We want to describe the problem in our requirements. So, this is what I'm trying to do; I need to steer, I need to brake, I need to make it go. These are the way that we want to get our User Stories. Not "give me an engine." That's not what we're looking for.

Let's move on; let's talk about satisfiers. Satisfiers, these are the things that you really kind of expect in your purchase. In a car, I would likely expect cup holders. I would expect windshield wipers. I would expect comfortable seating. Now, let's go for this example, and as we talked about priority being or value being subjective. For me, heated seats are a must-have. I will not buy a car without heated seats. I know it sounds silly, but I live in the south. I live in South Carolina, but I like heated seats, so I would not buy a car if it doesn't have heated seats. For me, that's a must-have. I need to have that or I'm not going to buy this car. So, you get the idea here. These are the things that are the basic categories.

And delighters, that could be self-driving or satellite radio, those kinds of things. If you don't get them, you're still going to be okay, but if you could get it, that would be awesome, that would be really cool and make you really excited about your purchase. And these are some techniques to help you categorize.

Categorizing using MoSCoW or Kano is only the beginning. The next essential step is to prioritize within these categories, especially within the 'Must Have' segment. This can be based on various criteria such as value proposition, urgency, or dependencies.

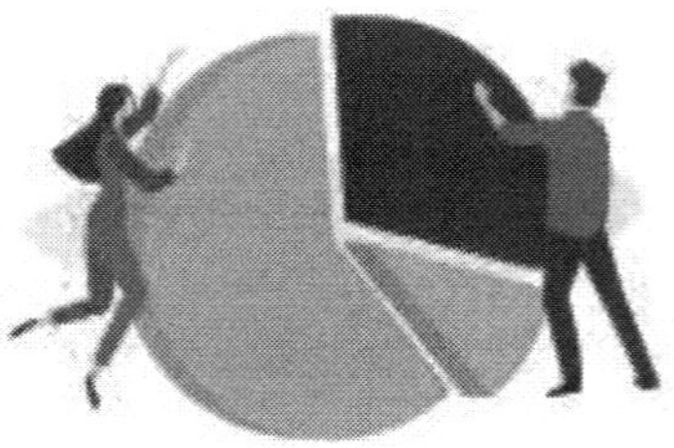

DEALING WITH RISK

Agile approaches come with a suite of built-in mechanisms designed to enhance visibility, encourage frequent feedback, and manage risk. The likes of Sprint Review, Prioritized Backlog, and Daily Scrum are native to agile practices, ensuring that the team has a structure to identify and address concerns in real-time.

The Common Mistake: Relying solely on these built-in structures can be a pitfall. Many Agile teams, engrossed in the rhythm of sprints and releases, often overlook broader risk management outside of these mechanisms. Ignoring intentional, dedicated risk management can lead to unforeseen problems later on.

Decentralized Decision-making:

1. **The Misconception**: A prevalent misconception is that a single decision-maker leads to clarity and reduces chaos. In reality, this can result in bottlenecks, reduced availability, and potentially myopic decision-making.
2. **The Agile Perspective**: Agile promotes decentralized decision-making where team members, being closest to the ground realities, are empowered to make crucial decisions. It's about tapping into the collective intelligence of the team, which often results in more informed, timely, and effective decision-making.
3. **Transparency**: One of the pillars of Agile, transparency ensures that even as decisions are decentralized, everything remains visible. From Story Boards to daily

discussions, everything is documented and open for scrutiny. This continuous visibility ensures that the team's direction aligns with the project's goals and stakeholders' expectations.

4. **Mentorship Over Dictatorship**: Transforming a single decision-maker from a dictator to a mentor can be a game-changer. Instead of providing directives, they guide the team with insights, helping them navigate challenges and make better-informed decisions.

Risk Reduction Through Decentralization: Having multiple heads involved in decision-making can be more effective in identifying potential pitfalls or missed opportunities. As the team collaborates, validates, and makes collective decisions, it invariably taps into a wider knowledge base, reducing risks associated with unilateral decisions.

Risk management in Agile isn't just about the built-in processes; it's about creating an ecosystem where every team member feels empowered to raise concerns, make decisions, and ensure the project's success. Decentralized decision-making, backed by the bedrock of transparency, paves the way for a more resilient and adaptive project management approach.

RISK AS A FACTOR OF PRIORITIZATION

When considering project management, particularly within the Agile framework, risk stands as a crucial element that significantly impacts the prioritization process. This approach isn't exclusively for Agile methods; integrating risk-based prioritization can significantly improve even traditional project management practices.

It's just a model to use. You don't have to follow this to consider yourselves Agile. but it might help. I would suggest you could probably do this as part of your Traditional projects as well. It'll help you have more successful Traditional projects. We start with high-risk high-value stuff. We always focus on value first and then focus on risk second. High value first, high risk is our second attribute of prioritization.

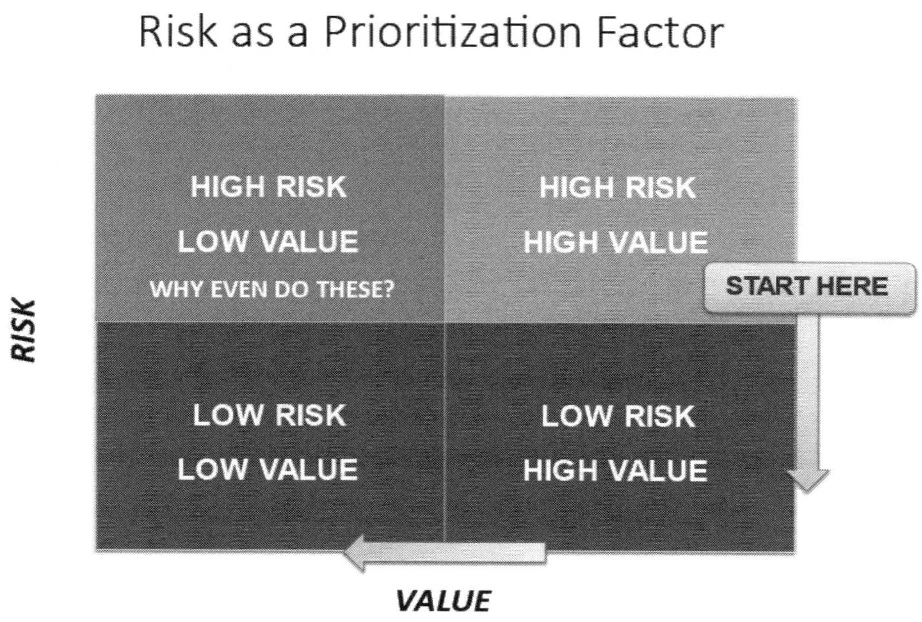

Figure 16: Risk as a Prioritization Factor

1. High-Value, High-Risk: Tackling high-value, high-risk items first makes logical sense. These are tasks that offer tremendous benefits but come with uncertainties. By addressing these elements early in the project cycle, you can mitigate significant risks upfront. This approach ensures that the challenges keeping you awake at night are dealt with promptly, leading to a smoother project journey in its latter stages.

2. High-Value, Low-Risk: Once the most challenging aspects have been addressed, shift focus to tasks that promise significant value but are associated with minimal risks.

With most uncertainties already out of the way, this phase should progress relatively smoothly, offering considerable benefits.

3. Low-Value, Low-Risk: By this phase, you'd have accomplished the project's primary objectives. What remains are tasks that, while not essential, can enhance the final product without introducing substantial risks. These are the cherries on top, and while they might seem like simple add-ons, they can often enhance the end user's experience. At this stage, teams might evaluate the importance of these tasks and decide whether the effort required aligns with the value they provide.

4. High-Risk, Low-Value: These tasks are tricky. They present significant challenges without offering substantial value. More often than not, these tasks may not be worth the effort and resources they demand. It would be prudent for teams to evaluate if these tasks are truly essential or if they can be eliminated from the project scope without negatively impacting the end product.

Concluding Thoughts: Risk-based prioritization ensures that your project is always geared towards delivering maximum value while effectively managing uncertainties. It encourages a proactive approach, addressing challenges early on, and ensuring that as the project progresses, the team is primarily left with tasks that are less daunting and more straightforward. This strategy not only increases the chances of project success but also ensures that resources are utilized in the most efficient manner possible.

Now, you might ask "why do we want that risk stuff at the top?" Well, think about it this way, wouldn't you like to get that big scary stuff that keeps you up at night out of the way as early as possible on a project? And then, by the time you get to the end of your project, that scary stuff is done, and we're left with things to do with very little risk. That's a really healthy place to be on a project. Get that scary stuff out of the way as early as you possibly can, go do the prototyping and research you need and get that risk mitigated fast.

DETERMINING THE RISK SCORE (OR FACTOR)

The essence of risk management in any project is to ensure that uncertainties do not derail its primary objectives. By identifying, analyzing, and responding to risks, a project can better navigate challenges and ensure a smoother progression towards its goals. We do this by determining a risk score, for example, on a scale of 1-3, find the risk probability based on expert judgment and also determine an impact or severity (again based on expert judgment). Multiply both and you have what we will call a risk score or a risk factor.

Example: Risk of employee being late for a critical regulatory meeting:

Probability: 1 (it holds at noon, employee would be at work)

Impact: 3 (stiff penalties are imposed for lateness or no-shows)

Risk Score (or Factor) = Probability Rating x Impact Rating

Risk Score = 1 x 3 = 3

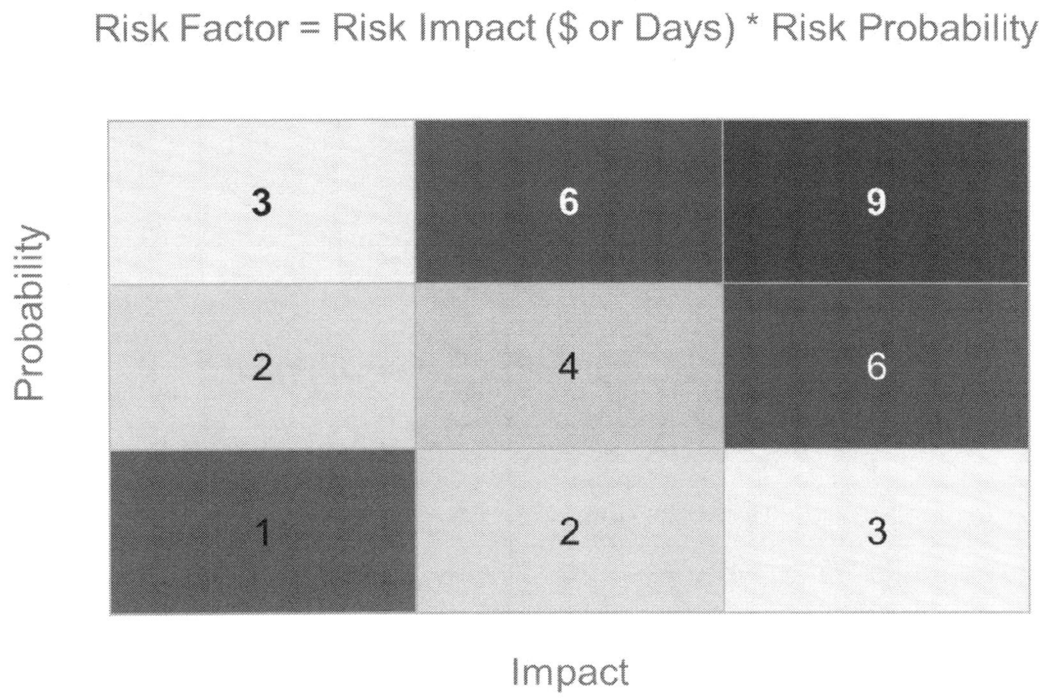

Figure 17: The Concept of Risk in Agile

1. **Risk Factor using Impact and Probability**:
    - **Risk Impact**: This denotes the potential damage a risk can cause if it materializes. It can be measured in terms of financial loss, delays (days), resource drain, or any other relevant metrics.
    - **Probability**: Represents the likelihood of the risk event occurring. This is usually expressed as a percentage.
    - **Risk Factor Calculation**: The standard way to compute the risk factor is by multiplying the Risk Impact and Probability. This yields a numeric value that can help in prioritizing risks.
    - **Formula**: *Risk Factor = Risk Probability x Impact*

2. **Risk Chart**:
    - A more visual and intuitive approach involves plotting risks on a two-dimensional chart, with one axis representing Impact and the other Probability.
    - Risks are then categorized based on their positions:
        - **Low Impact, Low Probability**: Assigned a risk score of 1.
        - **High Impact, High Probability**: Could have a score of 9.
    - The chart offers a clear visual representation, making it easier to understand the relative significance of each risk.
3. **Risk Threshold:** Once risks are plotted or their risk factors calculated, the next step is to determine thresholds for mitigation action (risk spikes). Risks below a certain threshold might be considered tolerable and may not necessitate immediate mitigation strategies. Conversely, risks that surpass the threshold are flagged for urgent action. For instance, risks with a factor below 4 might be considered manageable without aggressive intervention. Those with a factor of 4 and above would be prioritized for mitigation, given their potential to significantly impact the project.

Regardless of the method employed, the primary objective remains to quantify and categorize risks. This facilitates informed decision-making, ensuring that the project remains on track and resources are channeled appropriately. Integrating such a structured approach to risk management within Agile methodologies ensures that teams are always prepared to tackle challenges head-on.

RISK ADJUSTED BACKLOG

There's something else we need to do. It's not enough to just identify your risks. You've got to do something. The way that we do that in Agile is this; we do a Risk Adjusted Backlog. We take those risks that we've identified, and we calculate the risk score (factor), and then based on our risk threshold, we identify whether we need to take action or not. Then from there, we decide what actions we want to take to mitigate these risks. Those actions will likely take the form of a Spike in our backlog.

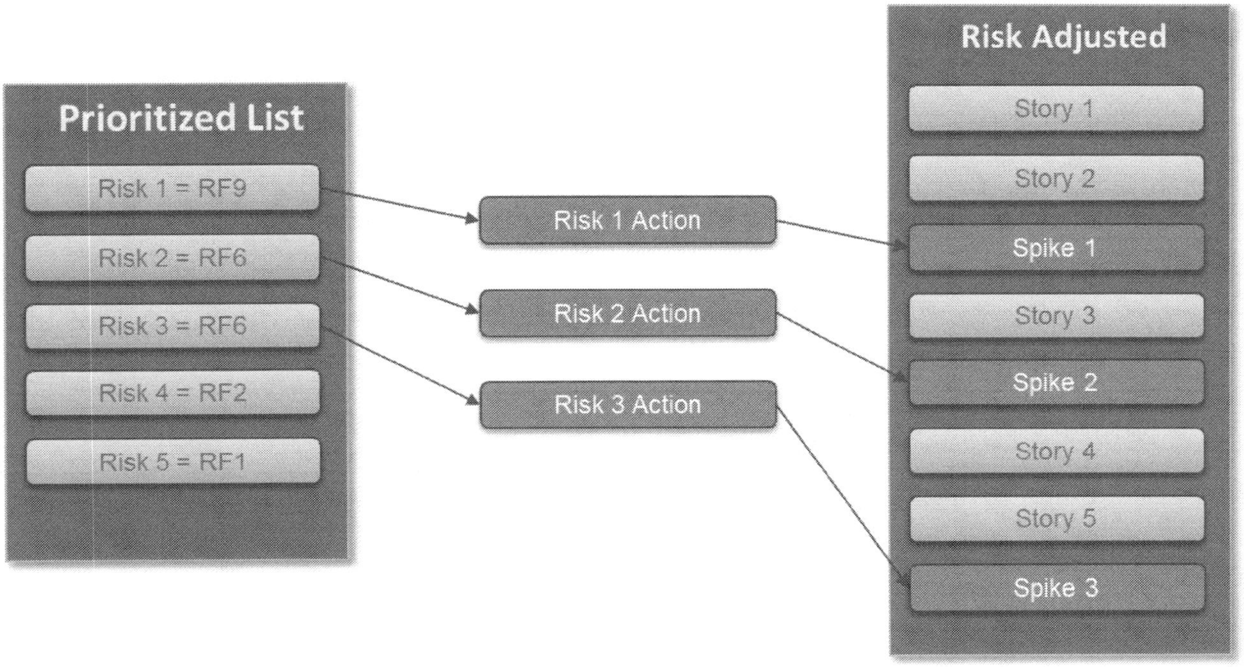

Figure 18: Concept of the Risk Adjusted Backlog

Understanding the Risk Adjusted Backlog:

1. **Risk Identification and Analysis**: Start by listing all possible risks. These could range from technical challenges to personnel issues to stakeholder engagement. Each risk is then analyzed, using tools like the Risk Score or Factor we discussed earlier.

2. **Actionable Risks**: Based on the threshold set during the risk analysis phase, certain risks are flagged as requiring action. These are the ones that could, if they materialized, significantly derail or harm the project.
3. **Risk Mitigation as Work Items**: The actionable steps needed to mitigate these identified risks are transformed into Spikes or user stories. A Spike, in Agile terminology, is a task aimed at answering a question or gathering information, rather than producing shippable product.
4. **Integrating into the Backlog**: These risk-mitigation items are added to the project backlog. They are treated like any other task, with estimations of effort, prioritization, and assignment. This ensures they aren't lost or forgotten but are instead an integral part of the project's work flow.
5. **Prioritization**: Like all other items on the backlog, risk-mitigation tasks are prioritized based on their importance and urgency. This ensures that high-risk factors are addressed promptly.
6. **Tracking and Review**: As with all tasks in the backlog, progress on risk-mitigation tasks is tracked and reviewed. This ensures that risks are not only identified but also actively managed.

Benefits:

1. **Visibility and Transparency**: By integrating risk management directly into the backlog, all stakeholders, from team members to product owners, have visibility into the risks and the efforts to address them.
2. **Better Resource Allocation**: Since risks can take up team capacity, incorporating them into the backlog helps in effective sprint planning, ensuring the team isn't overburdened.
3. **Proactive Risk Management**: Instead of reacting to risks when they manifest, the team is taking proactive steps to prevent or minimize them.

Conclusion:

The Risk Adjusted Backlog is a tool that embodies the Agile principle of adapting to challenges. It ensures that risk management is not a sidelined activity but an integrated part of the Agile project lifecycle. This proactive approach to risk enhances project predictability and delivers better outcomes.

QUESTION 26

MoSCoW stands for __________

A. Might Have - Should Have - Could Have - Would Like
B. Must Have - Should Have - Could Have - Will Love
C. Must Have - Should Have - Could Have - Would Like
D. Might Hope - Shall Have - Could Have - Will Like

So, the first level, you can imagine, it has to be the must-have. Those things that you absolutely have to have. If you look at a might have it's really kind of like should have or could have. Very, very similar, so it should be those baselines must-have requirements, the should-have requirements, the could-haves, and then the dessert toppings (would likes).

Answer: C. Must Have - Should Have - Could Have - Would Like

QUESTION 8

Kano Analysis consists of:

A. Must Have – Satisfiers – Delighters – Dissatisfiers
B. Might Have – Satisfiers – Disdainers - Duties
C. Must Have – Strengths – Delighters - Dissatisfiers
D. Might Hope – Satisfiers – Delighters - Dampers

Using the process of elimination, you could strike out some bogus options such as duties, strengths, and dampers. The answer is therefore A.

Answer: Must Have – Satisfiers – Delighters – Dissatisfiers

CHAPTER 7

STAKEHOLDER ENGAGEMENT & TRANSPARENCY

ENGAGING STAKEHOLDERS/TRANSPARENCY

Engaging stakeholders in Agile is reminiscent of traditional project management. While the foundational principles remain consistent, Agile offers its own set of tools and techniques to foster this involvement. Central to this engagement is the principle of transparency. This emphasis on clear and open communication is a hallmark of Agile, ensuring that all parties, from developers to stakeholders, are on the same page.

Transparency is not just a buzzword in Agile; it's a lifeline. Without a clear view of the project's status, challenges, and progress, stakeholders are left in the dark. And when stakeholders lack visibility, it creates a vacuum. This void can lead to a myriad of issues, from mistrust to inefficient decision-making processes. It's not just about delivering periodic status reports or progress updates, but about providing continuous insight into the heartbeat of the project.

For managers, this visibility is paramount. When a manager is accountable for product delivery, they need to be equipped with accurate and timely information. A manager operating without clarity may inadvertently resort to undesirable practices, such as micromanagement, which can stifle a team's productivity and innovation. The aim isn't to bind managers' hands but to empower them with transparency so they can trust their teams to do what they do best.

This responsibility is dual-faceted. While managers need to foster an environment where teams feel trusted, teams, on their end, need to ensure that they provide this essential visibility. Without this, there's a risk that managers might feel the need to interrupt the team's workflow to extract status reports and updates. Such interruptions can be disruptive, slowing down the momentum and derailing focus.

But here's the beauty of Agile: it inherently promotes transparency. Agile frameworks, with their boards and tools, serve as windows into the real-time status of projects. Whether it's a Kanban board showcasing task progress or a Scrum board reflecting sprint statuses, these tools make hiding or obscuring project details challenging. They're designed to shine a light on every nook and cranny of a project, ensuring everyone, from team members to stakeholders, has a clear picture of where things stand.

In essence, the symbiosis between stakeholder engagement and transparency in Agile is profound. They reinforce each other, ensuring that projects not only move forward efficiently but also with the collective buy-in of all involved.

So again, transparency is absolutely critical, and the good thing is Agile builds transparency in. That doesn't mean that the team has no choice but to be transparent. They could hide things if they really wanted to, but it's a lot harder. We have the boards; we have a lot of good stuff to give us transparency.

- **Roadmap**
    - Visual for what is currently being worked on

- Shows dependencies across product/business areas
- Compare actual capacity against desired work

- **Short Iterations**
  - Keeps stakeholders from losing interest
  - Reduces risks and allows for frequent decisions
- **Sprint Review**
  - Builds trust
  - Opportunity to accept new ideas and feedback
- **Story Boards**
  - Visibility into everything that is currently being worked on, done and not started
  - Who owns what, anything blocked, current status
- **Reporting**
  - Burn charts
  - Metrics

Let's talk about some of these techniques, so the first is the **Roadmap**. That is a form of transparency. It is a visual indicator of what's currently being worked on and what's coming up in the future. It's our plan for what's going to happen soon. We'll talk more about the Roadmap in a little bit. You can show dependencies across areas, about across silos, things like that. And most importantly, we can also compare actual capacity against what is desired, so there might be something requested. Still, we may not have the capacity to actually get it done in the time that's being requested. We do some comparisons on these charts on this Roadmap to say, "here's what's being asked for, and this is the capacity needed, and here's the capacity available." That's a very useful thing to help us plan better.

Short iterations offer clarity and consistent stakeholder engagement. Delivering quickly can captivate stakeholders. However, if outputs aren't riveting or apparent, stakeholder interest might wane. It could be strategic to demo every other iteration, ensuring relevance and sustained interest. The goal is flexibility to maximize engagement, leading to enhanced ideas, trust, and reduced reporting. Sprint Reviews, offering demos, are critical to foster trust and garner feedback.

Story Boards: These visuals offer real-time insights into ongoing work, whether in a sprint or in-progress tasks in Kanban. They depict tasks, their statuses, and owners. If stakeholders wish to gauge progress, they can glance at the board, saving the hassle of generating status reports. Details like vacation plans of team members can be inferred, enabling proactive problem-solving. Story Boards, though intrinsic to Agile, can also be integrated into traditional methods.

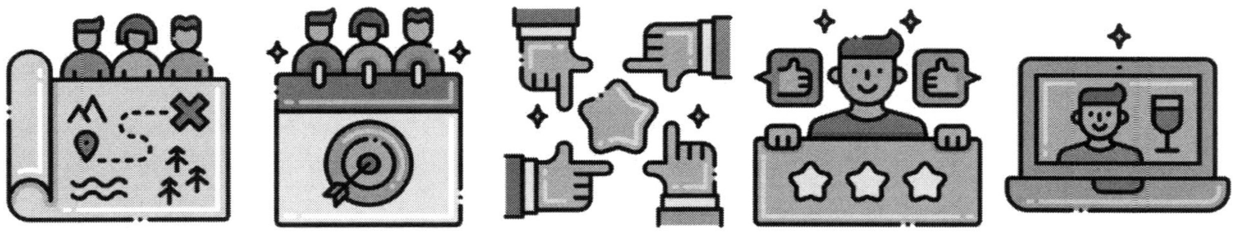

Reporting & Metrics in Agile: Agile thrives on data, utilizing tools like Burn charts. From Burndown to Burnup charts, a plethora of data is at one's disposal. Every team activity is monitored, facilitating detailed reporting on durations, tasks, and defects. However, this data-rich environment poses challenges. It's imperative to use metrics constructively, aiding team progress rather than as punitive measures. For instance, Velocity in Scrum denotes the average team output. Fluctuations in Velocity aren't necessarily negative; they could arise from varied factors like team availability. Metrics should prompt supportive inquiries rather than accusatory reactions. We'll venture deeper into metrics subsequently.

PRODUCT ROADMAP

You can craft roadmaps in myriad ways. What suits you is ideal. Some tools can assist in roadmap creation.

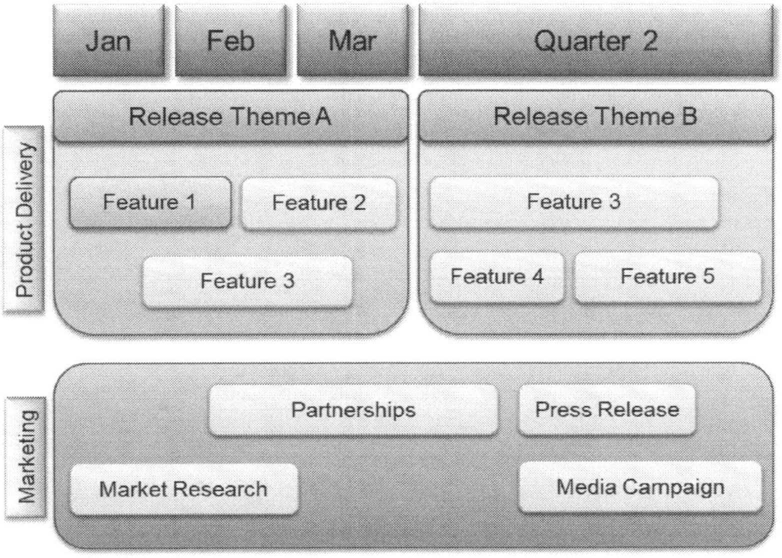

Figure 19: Product Roadmap Example

Consider this example: it offers a basic glimpse of features slated for implementation across a timeline. Understand that a roadmap represents a plan, not an unwavering commitment. The current quarter may have clearer deliverables, but as we move to the next, clarity may blur due to changes or reprioritization.

On your roadmaps, incorporating multiple swimlanes can be beneficial. It aids in identifying cross-swimlane impacts. For instance, one swimlane might focus on product delivery, another on marketing activities related to a product launch, and yet another on sales events or conferences. All these can influence product delivery. Hence, placing them on your roadmap offers a comprehensive visual of possible impacts. A roadmap, in essence, is a simple, visual representation of upcoming activities. You might also want to include a swimlane indicating the needed capacity versus available capacity, especially vital as you proceed.

A roadmap is a plan, not a binding commitment. This concept might unsettle some, especially those from the project management sphere accustomed to timelines as set commitments. Agile appreciates deadlines and fixed time boxes, but the flexibility to adjust scope to meet those deadlines is essential. We shouldn't compromise on quality or rely on overtime. The aim is to manage scope first. As we gather insights, plans might shuffle, priorities might shift, or timelines might advance or delay. Roadmaps, therefore, should be fluid.

A practical approach is to break the current quarter down into months for clarity, while future quarters can be presented as a whole. The rationale is that our foresight becomes hazier the further we project. To emphasize this, you might overlay a gradient or shading on your roadmap, reinforcing that predictions become less precise over time. This method helps manage stakeholder expectations, highlighting that while current plans are solid, future ones are less definite.

Swimlanes on roadmaps can illuminate potential challenges. For instance, while the top swimlane in our example showcases what we're building, the lower ones could represent other influential activities like marketing events or other projects. Additionally, charting capacity needs against available capacity can be insightful, ensuring alignment between what's demanded and what's feasible.

A crucial aspect not previously discussed is roadmap ownership. Contrary to traditional thought, in Agile, those requesting work (like Product Owners or the business community) own the roadmap, not the developers. Typically, the requesting party provides the timeline, and the development team indicates feasibility. This dynamic, while seemingly unconventional, promotes a collaborative approach. While the business community has ownership of the roadmap and its timeline, the building team provides input on what's achievable, fostering a more accurate, albeit still fluid, roadmap through collaboration.

STORY BOARD

Here's an example of a Story Board. Story Boards are prevalent in both Scrum and Kanban, and in Scrumban, which merges the two. The primary purpose of a Story Board is to visually represent the work in its current state, detailing the tasks at hand and those forthcoming. By "up next," I'm referring to tasks due in the upcoming days, not immediately. The components of a Story Board are largely customizable; it's sometimes termed a Kanban Board.

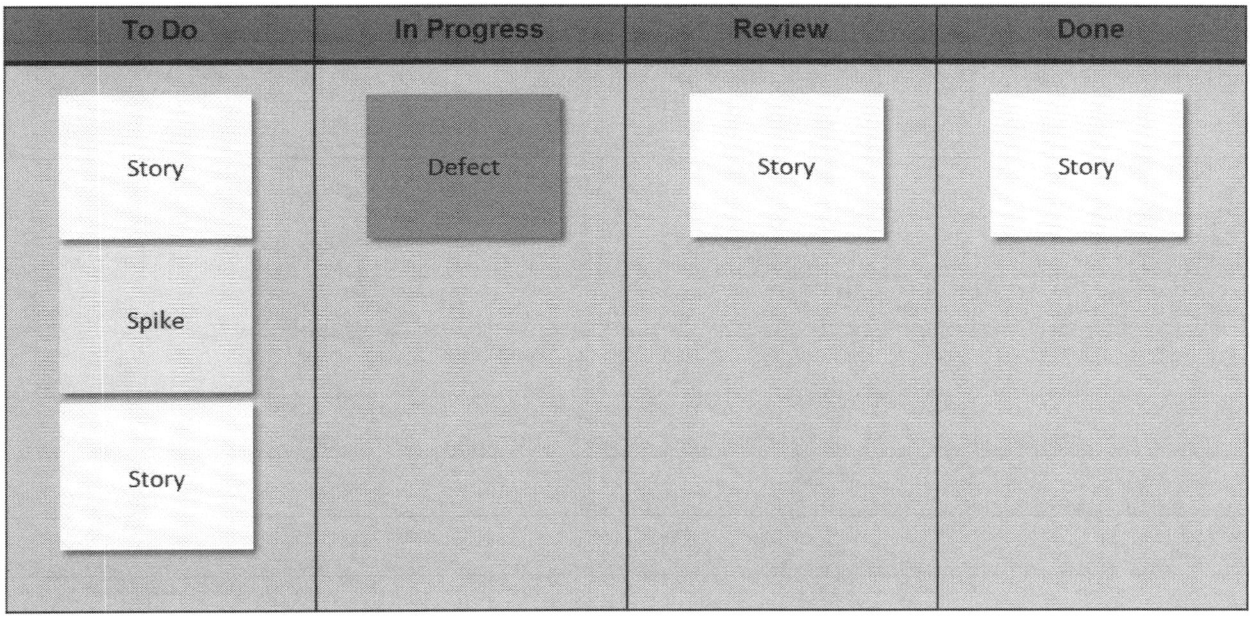

Figure 20: Story Board Example

The states of a Story Board are entirely configurable, based on your preferences. While it's possible to have multiple states, I personally prefer starting with a basic setup and evolving based on needs. It's best to avoid adding complexity until it's necessary.

The board typically displays categories like "to-do", "in progress", "review", and "done". The "to-do" list serves as a queue for upcoming tasks – items the team hasn't commenced yet. Meanwhile, "in-progress" showcases tasks that are currently underway. Often, these tasks are visually represented by cards, which is why some teams use actual whiteboards with Post-it® notes, sliding the notes across as tasks progress.

Popular tools, like Jira, offer such functionality. The primary goal is maintaining transparency, ensuring an accessible status of tasks at all times. This is the essence of a Story Board.

Many tools, including Jira, accumulate data by tracking the movement of these tasks (or cards) across the board. For instance, moving a task from "to-do" to "in progress" logs its start date, providing automated tracking. Once the task is complete and moved to "review", the tool calculates its duration in the "in-progress" state. This data gives insights into task durations, aiding team improvements.

At any moment, anyone can view the board to understand the team's activities and challenges. Visual cues, such as color changes, can highlight blockages. Furthermore, with electronic tools like Jira, you can venture deeper into tasks, accessing details like comments, descriptions, and more. Each card is essentially a reservoir of task-related information – from comments to audit trails. This visual representation provides an instantaneous snapshot of team operations, but there's always the option to investigate further.

The board functions simply: a task is pulled from "to-do" to "in progress", worked upon, and then moved to the next state until it's marked "done". Tools like Jira and many others, through the mere act of shifting cards, collect valuable data – entry and exit times from states, ownership, transitions, and more. This data, far from being punitive, is instrumental for team improvements. For example, if a task lingers in "in progress" for an extended period, it warrants discussion.

Tools like Jira can be configured to display dots under a card, representing the days it has been in its current state. An accumulation of dots signals potential issues, prompting conversation. Remember, this data isn't for reprimanding but for initiating constructive dialogues.
Agile methodologies, through these visual tools, don't directly resolve problems but highlight their existence. This clarity is the real strength of Agile. It alerts you to issues, leaving the rectification up to you. The primary goal is to act promptly upon noticing discrepancies. All these tools and metrics aim to provide the clarity and data necessary to facilitate improvements.

That's the intent and the general idea. All these things are to give us the transparency and the data we need to go off and go fix stuff!

CHAPTER 8

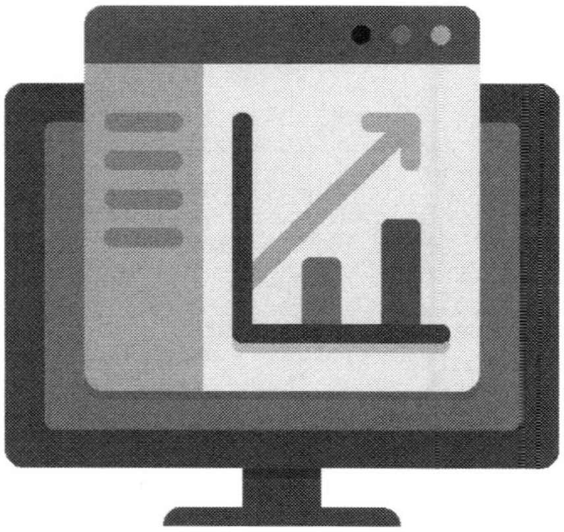

AGILE METRICS

METRICS

Metrics play an indispensable role in Agile. While data collection is abundant, it's crucial to exercise caution. Data can be both a tool and a trap, and understanding its nuances is key.

1. **Vast Data Collection**: With modern tools, almost every activity is tracked. From task movements to ownership changes, everything is recorded. But remember, just because you have data, it doesn't mean it provides insight.
2. **Potential for Misuse**: Data, if not understood or interpreted correctly, can mislead. It's not just about the numbers; it's about understanding the story behind them. Engage in meaningful discussions around metrics rather than drawing quick conclusions.
3. **Purpose-Driven Metrics**: Different metrics serve different needs. Some forecast future trends, some track present progress, and others identify areas of

improvement. It's essential to use them appropriately, ensuring they bring about positive changes and aren't used as mere tools of judgment.

4. **The 5 Why's**: This lean concept is about understanding root causes. Instead of taking things at face value, it encourages probing deeper into problems by asking "why" multiple times. But, a word of caution: use it judiciously. While it can be a tool for clarity in professional scenarios, in personal ones – like the humorous example with the wife – it can backfire.

Popular Agile Metrics:

1. **Burndown & Burnup Charts**: The Burndown chart displays remaining work, while the Burnup chart shows work completed. Although they offer different perspectives, both are vital for understanding a project's progress. While Burndowns often reflect task completion, Burnups might give a clearer picture of work achievements.
2. **Completed vs. Committed**: A reflection of a team's predictability. It's the ratio of work completed in a sprint compared to what was initially committed. A team with consistent results close to their commitments is more predictable, making future planning and forecasting more accurate.
3. **Story Points**: These are indicative measures of work. At the start of a sprint, teams commit to a number of story points. By the end, they tally up what they've achieved. It's a way to gauge team efficiency and predictability.

Henrik Kniberg, a known figure in the Agile community, said, “It's not the process that's important, it's the process to improve your process.” In essence, Agile is about continuous improvement. Metrics are just tools in that journey, helping teams introspect, adapt, and evolve. Alongside other metrics like quality, they ensure that Agile teams stay on the path of progress and excellence.

BURNDOWN CHART

Let's talk about the Burndown chart and what that looks like. Burndown charts are a way of showing progress are we getting to completion. The Burndown chart is a visual representation of work progress in Agile methodologies. At its core, it is designed to provide insight into how close the team is to completing all their committed tasks within a sprint.

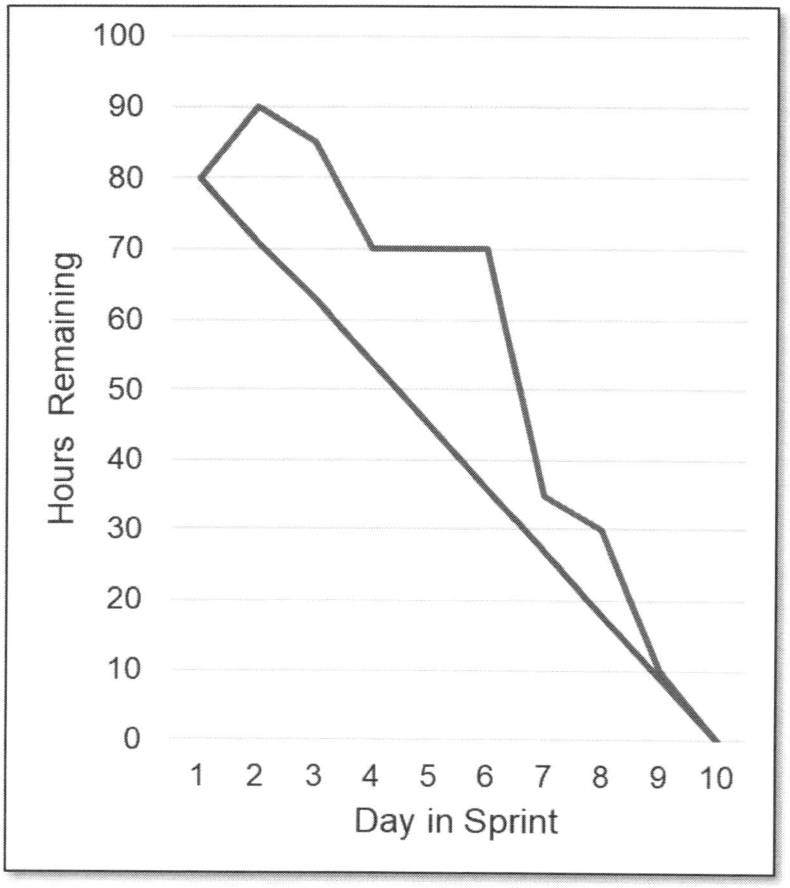

Figure 21: Burndown Chart Example

Components:

- **Y-Axis**: Represents the estimated hours required to complete tasks.
- **X-Axis**: Represents the days within a sprint. In our example, this is a 10-day (2-week) sprint.

Construction:

- At the onset, each task within the sprint backlog is estimated in terms of hours.
- These cumulative hours represent the workload for the entire sprint and are plotted as the starting point of the chart.

Sample Observations from Figure 21: Burndown Chart Example:

1. **Starting Point**: The team begins with 80 hours of work for the entire sprint. This initial estimate derives from the collective assessment of each task's effort.
2. **Day 2 Jump**: The total hours needed jump from 80 to 90. This can occur due to various reasons such as task underestimation, unforeseen complexities, or even the addition of new tasks. While this jump can be alarming, it's essential to understand its cause rather than jumping to conclusions. The team can discuss it during the daily standup to adjust and realign.
3. **Day 4-6 Flatline**: No work seems to have been done, as indicated by the flat section of the graph. Such plateaus could result from holidays, a team member's absence, or external blockages. It might also be a result of the team not updating their progress. Regularly updating progress is vital for the chart to remain a useful tool.
4. **Dramatic Drop at Day 6**: A sudden decrease in hours suggests a rapid completion of tasks. Various reasons can account for this—maybe some team members returned from a break, a blocker was removed, or several tasks got completed simultaneously. Such instances can be positive learning points for retrospectives.

The End Goal:

- The ideal trajectory of a Burndown chart is a steady downward slope, indicating a consistent pace of work.

- Significant deviations—be it flat lines, spikes, or sharp drops—act as triggers for team discussions. These deviations signal that there may be issues or opportunities the team should address.

The Burndown chart serves not as an evaluation tool, but as a conversation starter. It helps the team reflect, adjust, and strategize. Recognizing patterns, understanding deviations, and adapting accordingly are the key takeaways from this metric.

QUESTION 27

Velocity represents _________

A. How much work the team can plan in one project management life-cycle

B. How much work the team can finish in one sprint

C. How much work the team can commit to in one project backlog

D. How much work the team can finish in one daily scrum

I just want to put the one comment out there about velocity, and we'll talk more about it soon. It's a great Metric to use for forecasting purposes. That's the only reason that we use velocity, just to forecast. Never ever, ever use it for performance metric. I'm going to say that a few times today, just to keep that cemented. Do not use it to compare teams, and there's lots of reasons for it. We'll discuss that a little bit deeper later.

Answer: How much work the team can finish in one sprint

QUESTION 28

In a burn-down chart, what is on the Y axis?

A. Story points

B. Sprints

C. Months

D. Weeks

The Y axis, the axis on the left are vertical axis is hours or Story Points. You can do either one. In this case with this example, it will be Story Points. So it'll be Story Points or level of effort, consider it that way. Level of effort on the left-hand side, in the Y axis and on the bottom the X axis is the days in the sprint.

BURNUP CHARTS

We're going to be talking about **Burnup charts**, which is the corollary to the Burndown. Burnup charts, in essence, reflect progress by showcasing work that has been completed over time. Instead of diminishing work, like in a Burndown chart, you're accumulating completed work in a Burnup chart.

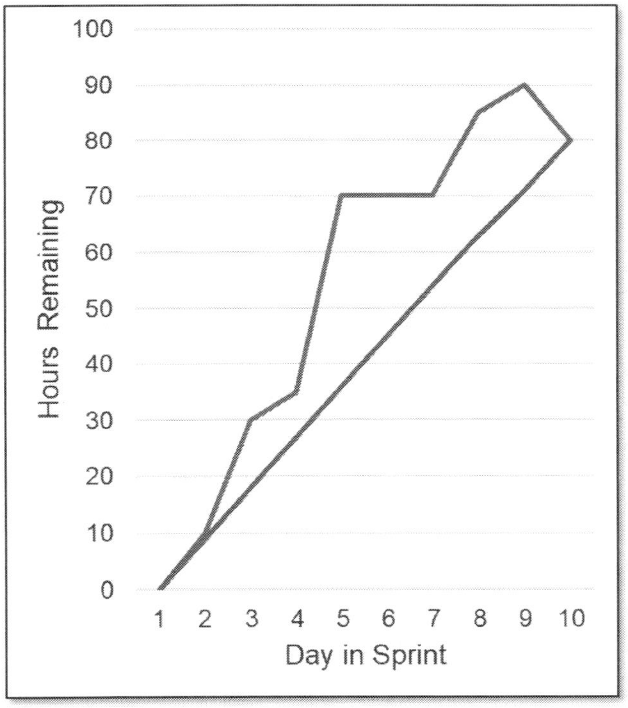

Figure 22: Burn Up Chart Example

All right, let's jump further. Okay now, here's another example of a Burnup chart.

Now, this second chart is also a Burnup chart but in this case it spans across a series of sprints and a longer timeline. You can create a burn chart for a sprint, you can do it for an epic or a feature or release. It's really just level of effort on the left and the days in the time box on the bottom.

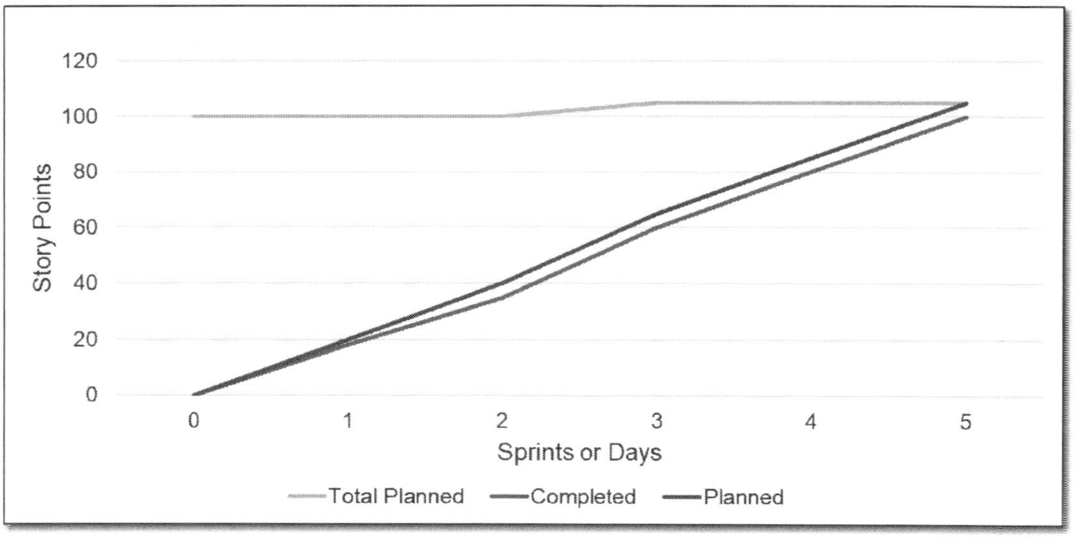

Figure 23: Release Burn Up Chart Example

Components:

- **Y-Axis**: Represents the cumulative effort (in terms of story points, hours, etc.)
- **X-Axis**: Denotes time, which could be days in a sprint, sprints in a release, etc.

Construction:

- **Yellow Line (Total Scope)**: Signifies the total anticipated work for a given time box. It represents the current understanding of all the work required to achieve the objectives.
- **Red Line (Planned Progress)**: This line symbolizes the ideal trajectory of the work if everything proceeds as anticipated.
- **Green Line (Actual Progress)**: This shows the real-time accumulation of completed work over the time box.

Key Takeaways from Burn Up Chart Example:

1. **Beginning**: The release starts with an anticipation of 100 story points worth of work to be done. This is your initial total scope.
2. **Scope Change**: Around sprint 2-3, the yellow line jumps, signaling an increase in scope. Such an occurrence prompts essential dialogue within the team about feasibility and potential changes to the release or sprint's objectives.
3. **Adjusting the Plan**: Post scope increase, the red line (plan) adjusts, projecting the new path the team should ideally follow.
4. **Completion**: The green line trails behind the red line, suggesting the team couldn't entirely meet the planned progress by the end of the time box.

Benefits:

- **Visibility**: Clearly visualizes scope changes, aiding stakeholders in understanding the real-time impacts of added requirements.
- **Ease of Use**: The team's primary action is moving stories to the "done" column, automating the chart's update process.
- **Transparency**: Highlights gaps between planned and actual progress, offering insights into potential roadblocks or interruptions the team faces.

Points to Remember:

- Agile doesn't advocate for capturing every minor detail upfront, especially for longer timelines like epics. The process should be iterative, capturing just enough information to move forward efficiently.
- While it's an excellent tool for tracking, it also serves as an indicator of scope changes and possible disruptions. This makes it instrumental for stakeholders and project managers to initiate proactive interventions if needed.

Conclusion: Burnup charts, by graphing completed work against time, serve as an invaluable tool for Agile teams. They enable better communication, showcase progress and changes in scope, and aid in anticipating and resolving challenges. Coupled with other Agile tools and practices, they contribute significantly to fostering transparency, clarity, and adaptability in projects.

CHAPTER 9

AGILE TEAMS AND COLLABORATION

BUILDING AGILE TEAMS

When constructing teams in the realm of Agile, the methodology and mindset differ considerably from traditional approaches. Agile places significant emphasis on team dynamics, emphasizing intentionality in team building.

Agile teams are typically small. As previously discussed, while the Scrum Guide recommends teams of 3-10 members, it's common to see slightly larger teams. However, the larger the team, the greater the complexity and potential challenges. Thus, teams of around 9 or 10 are often optimal, and teams of 12 might be pushing the envelope. The directive is clear: **build small teams**.

A common concern that arises is the need for a larger workforce on extensive projects with tight timelines. The solution in Agile isn't to inflate a single team's size but rather to create multiple teams. Instead of overloading one team, construct several, ensuring they have opportunities for communication and collaboration. The rationale behind this is straightforward: smaller teams reduce complexity, facilitating better communication and bonding.

Agile teams should also be diverse in terms of skills and viewpoints. A team rich in complementary skills and diverse perspectives can tackle challenges more effectively. A vital component is the ability of team members to challenge each other constructively, promoting innovation and better solutions. This challenges the traditional notion of the lone wolf; in Agile, the focus is on collective performance.

Another cornerstone of Agile teams is commitment. They are devoted to a shared purpose. It's not merely about individual achievements but holding every team member accountable. Peer accountability is potent—team members are more hesitant to disappoint their peers than their superiors. This interdependency is a strength, driven by the human desire not to let one's peers down.

However, for this system to function, a high level of trust within the team is paramount. A team member pointing out another's oversight isn't seen as a criticism but as a means of ensuring collective progress. Such interactions, if not based on trust, can lead to conflicts. Building such trusting teams is challenging but essential for Agile's success.

T-shaped skills or what's known as "generalized specialists" is another pivotal element in Agile teams. The idea isn't to transform everyone into a jack-of-all-trades. Rather, the goal is to have team members who excel in their primary skill but can also contribute in other areas. Such versatility ensures that team members can support each other, promoting skill transfer and mentorship. This approach is beneficial not only for the team's development but also for its resilience. If a key team member is unavailable, work doesn't come to a halt. Teams equipped with complementary skills can continue to operate efficiently, emphasizing the Agile value of adaptability.

In conclusion, building Agile teams is more than just grouping individuals. It's an intricate process, centered around trust, commitment, diversity, and versatility. Investing time and effort into building such teams is the foundation of Agile's success.

TEAM FORMATION – TUCKMAN MODEL

Now, this is just one model that that you can make use of. It's called the Tuckman Model or Tuckman's Ladder. You're probably familiar with this. This is called **forming, storming, norming, and performing**. All teams go through a similar flow of forming a team. This is a pattern that we typically see in their maturity.

Figure 24: Tuckman Model (5 Stages of Team Development)

Tuckman's model for team development identifies five stages of team development which project teams must go through before the team can achieve maximum effectiveness. Tuckman's Stages of Team Development are *Forming, Storming, Norming, Performing* and *Adjourning*. As a team evolves, the team consciously or subconsciously moves from one stage to the next. Due to changing circumstances, teams may go through the development cycle several times (e.g., a team in the performing stage could revert to the storming stage due to a change in team members).

1. *Forming* – In this first stage of team development, the team is formed. The team meets for the first time, members get to know one another and learn about the project. Team members act independently at this stage and are on their best behavior.
2. *Storming* – This second stage in team development is the most challenging. As project management approach, project work and technical decisions are addressed by the team, team members offer opposing ideas about roles, responsibilities, goals, and expectations. Conflicts may arise due to interpersonal differences. Teams may or may not move out of this stage. The more mature the team members are, the faster the team is able leave the storming stage.
3. *Norming* – Trust and motivation levels increase in this third stage as team members get more comfortable with each other and more familiar with the project.
4. *Performing* – At this stage, the team is high performing, functioning as a cohesive unit. Team members are interdependent, autonomous, and able to make decisions without supervision.
5. *Adjourning* – In this stage, the project is completed, and the team is disbanded. This stage is also known as the *mourning* stage due to uncertainty about the future and a sense of loss felt by team members.

There's actually another thing (not a formal stage) that can happen. It's called hyper norming and **hyper-norming is where you get to a stage of the team where everyone just agrees with everyone else**. They know each other, they trust each other, they just somebody has an idea, okay let's try it, you know. We want to encourage disagreement in order to get to that performing stage. If you're hyper norming, there's no disagreement. People are just doing whatever someone says. You might have some influencers that are making all the decisions and that's not a healthy place to be either. It's not a bad team but we need to start encouraging some disagreement and there again, your Scrum Master can bring some healthy disagreement to the team to encourage that mindset that it's okay to disagree, no one's going to get mad at each other, right? So watch out for those challenges, too.

The Tuckman Model, as discussed, presents the evolution of a team over time. It's imperative to understand that team dynamics play a crucial role in project success. Forming a cohesive team takes time, energy, and a deep understanding of individual differences and strengths.

1. Understand the Journey, Not Just the Destination:

While the aim is always to get teams to the performing stage, it's vital to remember that each phase has its unique challenges and lessons. Leaders should focus on the journey, ensuring each member grows and learns from every phase.

2. Anticipate Regression:

Changes, whether minor or significant, can trigger a regression in the team dynamics. Leaders should be prepared to tackle this by re-engaging, re-aligning, and building trust again. For instance, the introduction of a new team member can send a team back to the forming or storming stage.

3. Continuous Feedback:

Feedback mechanisms should be in place throughout the team's development. Regular retrospectives, one-on-ones, or feedback sessions help address issues proactively. These mechanisms also foster a culture of openness and growth.

4. Celebrate Milestones:

It's essential to celebrate when a team moves from one stage to the next. This not only motivates team members but also solidifies the bonds formed during the process.

5. Provide Necessary Training:

If a team seems stuck in the storming stage due to conflicts, consider conflict resolution training. Similarly, if the team struggles to communicate effectively during norming, communication workshops might be beneficial.

6. Leader's Role:

Leaders should be able to identify the team's current stage and tailor their approach accordingly. For instance, during the storming phase, a leader's intervention in conflict resolution might be necessary. However, in the performing phase, a more hands-off approach, letting the team navigate on its own, might be more beneficial.

7. Protect the Team:

Once a team reaches the performing stage, it's a gem that needs protection. Leaders should avoid drastic changes that can destabilize the team dynamics. If changes are unavoidable, they should be introduced gradually, with ample communication and support.

8. Recognize Individual Contributions:

While team dynamics is about the collective, individual contributions shouldn't be overlooked. Recognizing individual efforts ensures team members feel valued and encourages continued contribution.

The Tuckman Model isn't just a theoretical framework. It's a practical guide for leaders to understand, form, and guide teams to their maximum potential. Recognizing and navigating these stages effectively can mean the difference between project success and failure.

HIGH PERFORMING TEAMS

What are high-performing teams? These teams are self-organizing. They're not role title-based. They just do the job. Whatever is necessary they just do it! They're empowered to make their own decisions. This is that decentralized decision-making capability. The team has the ability to do their jobs. They know their jobs well. They can organize around it and they can make their own decisions about the best way to do the work. They have the information, so who better to make decisions?

So, we've understood the characteristics of high-performing teams. Let's venture a little deeper into the aspects that catalyze their efficiency:

1. **Shared Vision and Purpose:** For a team to perform at its peak, each member must have a clear understanding of the end goal. When the purpose aligns with individual values, the motivation is intrinsic, leading to passion and commitment.
2. **Diverse Skill Sets:** High-performing teams have a mix of experts across different domains. This diversity fosters innovation as members bring unique perspectives and expertise to the table.
3. **Open Communication:** Trust is the cornerstone of any team's success. For members to voice concerns, celebrate successes, or share innovative ideas, open communication is vital. This means creating an environment where all voices are heard and valued.

4. **Feedback and Reflection:** Continuous feedback, both positive and constructive, is the key to improvement. High-performing teams have regular check-ins to discuss what's working and what's not, ensuring constant alignment and improvement.
5. **Empowerment and Autonomy:** When individuals feel trusted and empowered to make decisions, they take ownership of outcomes. Empowering teams leads to faster decision-making and increased accountability.
6. **Growth Mindset:** High-performing teams embrace change. They understand that to stay ahead, they must continuously evolve. This means seeking out learning opportunities and being open to experimentation.
7. **Collaborative Culture:** While individual talents are recognized, these teams understand that collaboration is the path to success. They're willing to pool their skills and knowledge for the benefit of the larger goal.
8. **Effective Leadership:** Behind every high-performing team is an effective leader who understands how to inspire, guide, and provide the necessary resources. This leader is not an authoritarian but a coach who supports the team's journey.
9. **Recognition and Rewards:** Celebrating achievements, big or small, boosts morale. Recognizing team efforts and rewarding them fosters a sense of belonging and appreciation.
10. **Agile Mindset:** Flexibility and adaptability are essential. High-performing teams can pivot based on feedback and the ever-evolving external environment. They remain fluid and are not tied down by stringent processes.
11. **Healthy Work-Life Balance:** While commitment to the project is paramount, these teams understand the value of a healthy work-life balance. Ensuring team members have time for relaxation and personal commitments increases overall productivity and reduces burnout.

While it might seem that achieving a high-performing team status is an arduous task, the benefits are immense. From increased productivity to innovative solutions and a harmonious work environment, the investment in nurturing such teams is worth every effort. As you set forth in this journey, keep in mind that the journey to excellence is ongoing. Embrace the changes, nurture the team spirit, and cherish the successes.

SERVANT LEADERSHIP

This is a concept of how leadership works for the team and we're talking about servant leadership. It's not just for those who are labeled as leaders but also natural leadership as well. In Servant Leadership, we have four primary duties: **shielding the team from interruptions**, **remove impediments**, **communicate constant communication**, and what we call, **carrying food and water**. Get the team what they need so they can do their jobs.

Now, this is different from a Command-and-control approach, which is tell people what to do, Direct-and-control. That's not what we're trying to do. We don't want that single voice telling everyone what to do because then people can't make decisions on their own. If you want innovation, creativity and engagement, you have to back away and let the teams figure things out on their own. Your narrative captures the very essence of Servant Leadership. For leaders, it's not just a term; it's a journey, an evolution of self, and a way to empower and nurture their teams. Here's a breakdown of the critical elements highlighted:

1. **Focus on Serving:** Servant leaders are in the service of their teams. They're committed to supporting and empowering them, which results in increased team morale, creativity, and productivity. It's not about command; it's about cultivating a culture.

2. **Empowerment over Direction:** Instead of dictating tasks, servant leaders set a clear vision, and then step back, trusting the team to chart their course towards that vision.
3. **Shielding and Clearing the Path:** Servant leaders shield their teams from external distractions, removing obstacles, and ensuring that they have all the resources they need.
4. **Open Communication:** They maintain open channels of communication, always ready to lend an ear, advise, or provide clarity.
5. **Evolving from Traditional Leadership:** Transitioning from a command-and-control approach to servant leadership requires a paradigm shift. It's about recognizing the value in empowering teams and allowing them the space to innovate and grow.
6. **Recognizing One's Impact:** Leaders must be introspective, acknowledging that sometimes their very presence or involvement can be an impediment. The realization that 'less can be more' is vital.
7. **Questioning One's Role:** The dilemma of "If I'm not telling people what to do, what's my job?" is a turning point. Servant leadership is about facilitating, guiding, and serving – not directing.
8. **The Outward Focus:** While leaders should trust their teams to handle their tasks, they should focus on external factors like stakeholder management, creating partnerships, and setting strategic directions.
9. **Building Trust and Respect:** Ensuring teams are made up of skilled and trusted individuals, giving them the autonomy to operate, and entrusting them with responsibilities fosters mutual respect and trust.
10. **Personal Growth:** This leadership approach is not just about enhancing the team's performance but also about personal growth as a leader. It shapes you as a more holistic, understanding, and impactful leader.

11. **Continuous Learning:** The journey doesn't end. Leaders should seek continuous feedback, mentorship, and learning opportunities to evolve.

In essence, servant leadership is the embodiment of the phrase "leading from behind." It's a leadership style that prioritizes the growth, well-being, and success of the team above all. By embracing this style, leaders not only elevate their teams but also carve a path of personal and professional growth for themselves. It's a win-win and a testament to the power of empathy, trust, and empowerment in leadership.

It's a different approach, and quite honestly most of the organizations that I work with, leaders are uncomfortable with this style. Management certainly is because maybe they either haven't been trained or not exposed to this. It's not uncommon for our leadership to not understand this way of leading. Often, and I hear this a lot. When we encourage our managers to go this direction to be servant leaders and not tell people what to do, the biggest question that I get from those folks is "Well, if I'm not telling people what to do, what's my job?" That's a really interesting question when you really think about it.

Now, this is something that I went through myself, my own personal journey as a manager and as a leader. A number of years ago, I was a manager at a very large bank, international bank and I had an organization about 75 people, somewhere in this, a pretty good-sized group. And I had a number of Agile teams. Back then, this is early in my Agile days and I didn't really understand what it meant to be an Agile leader so I was kind of traditional and a little bit of command-and-control. I could freely admit that. I liked to be involved in the details and I liked to let people know that I was there, but I wasn't doing it the right way. I viewed it as my job as a leader to be there for my teams, so that meant going to all of their meetings. I was going to as many daily standups as I could. I was going to their planning sessions and their reviews. I was going to everything I could to be there for my teams. Well, I thought that was my job, right?

Fortunately, I had a really good Scrum Master that worked for me at the time, and keep in mind that she was 2 levels below my level and she had the courage...she trusted me enough that she came to me and she was a contractor also, which was even more impressive, she came to me and she said, "Roy I really appreciate what you're doing. I know that you're trying to help your teams, but you got to stop coming to my meetings. You got to stop." And I was a little taken aback by this. I was confused why she would say this because, you know, who is she to tell me that I shouldn't be going to these meetings. She said, "You're causing challenges and I'd like you to not come to my meetings and I think we can do better." I said, "Okay, so here's what we're going to do, I'm going to back away and I will give you 3 months to see how it goes. I'm going to stop coming to your meetings. You tell me which ones you want me to come to. I'm not disappearing. I'm always available, but I'm going to stop." So, I did that and I was really embarrassed because what happened is one month after me leaving, my teams got better.

They started producing more product, and I realized at that moment that I was the source of the problem by just being overly involved, being there in the meetings. It shut down the conversations because I was signing their paychecks. They didn't want to contradict me. They didn't want to show what problems we're having because while they trusted me, they didn't have that level of trust, right? That level of trust happens in the team and not necessarily with your management and your leaders.

That was my moment that I realized, that was my light bulb moment. I realized that I needed to change my behaviors in order to enable my teams to become great. I started questioning myself like well if I'm not telling people what to do, what's my job? I'm not doing that job anymore.

Fortunately, I have a mentor, someone that I've worked with for a number of years. He's the guy that taught me Agile many, many, many years ago. Very, very close friends and we still talk with each other often about different issues and things like that, doesn't mentor me as much anymore thankfully, right? I've gone past that a little bit, but we still bounce ideas off each other, so he knows how to coach and instead of telling me what to do. He started asking me questions like so what value do you think you can bring? How can you help your teams? And he started to ask me questions to help me get to the point where I could make a decision for myself, and I realized that instead of focusing downward on my teams, I need to focus outward. My teams, they're great people, smart, I hired well and that was a good job that I did. I hired great people. I always sure that we paid well and that we hired great people, and we treated them with respect, even my contractors. I made sure that we had great people so since I had set that up and I had great Scrum Masters, I didn't need to be as involved. Instead, I need to clear the path. I started focusing on external, right? The stakeholders.

I focused on my partners, and I found as a result, instead of focusing downward, I let the teams do their job. I set clear vision, gave them clarity, and started focusing more on stakeholder engagement. Clearing the path, reducing impediments, taking care of escalations and as a result of that, not only did my teams improved and their velocity dramatically improved, but my career got better too, because now instead of spending all of my time with my teams, I was spending most of my team with leaders, with people above me, right? I got more facetime and I got better known as a manager or leader that was getting results. This was a really hard thing for me to make that transition. I was a developer. I was writing software for years and for me to transition, which was really difficult and it took me a long time to get to that point, but that's what we're talking about. Changing that behavior, that structure and our personal behaviors to be those leaders that guide, that clarify, give vision and step away and let their teams do their jobs, and I'm promising you, that is a hard transition to make.

TEAM WORKING AGREEMENTS

Let's talk about team working agreements to enhance our teams even further. I always recommend teams to have a foundational agreement. This isn't a complex process. It's not about drafting an intricate contract. Rather, it's a simple consensus the team arrives at about how they wish to work in harmony. These are the fundamental principles and values that a team agrees upon.

Consider beginning with a Cadence Calendar. This is essentially a schedule of regular meetings or events the team will hold, such as the daily standup, planning sessions, reviews, and retrospectives. Discuss who should be present at these meetings. Who do we want to invite and who might we exclude? How should the team communicate?

As an example, I'm personally not keen on emails. Often, the essence of communication gets lost within lengthy email chains, leading to misinterpretations. After a while, the original purpose of the email gets buried under layers of replies, with countless individuals added to the chain, diluting the message even further. This is why I urge teams to have a consensus on their preferred mode of communication. If an email chain exceeds, say, three back-and-forth exchanges, it's perhaps time to halt it. Instead, pick up the phone, initiate a video call, or hold a face-to-face conversation. Once the discussion is over, record the conclusions. This prevents extended, unproductive email threads and facilitates faster, clearer communication.

It's equally crucial to have a shared understanding of conflict resolution. If disagreements arise, who should the team approach if they cannot reach a consensus? I'm also a strong proponent of setting meeting norms. Attending a team meeting? The team should concur that everyone will be fully engaged. This means no distractions, no checking emails intermittently, or doing other tasks. Being present and focused ensures meetings are productive and concise.

Let's be honest, many of us are guilty of multitasking. While I acknowledge that some of you might be doing other things right now, multitasking often detracts from the essence of a conversation. Hence, it's vital to be present, engage in the discussion, conclude it efficiently, and then move on. I also have a few rules for meetings.

I'm quite particular about punctuality in meetings. If the meeting starts at a set time, everyone should be there at that moment. Latecomers, especially for brief meetings like daily standups, need to provide an explanation. It's crucial to understand the reason behind the tardiness so that we can work towards eliminating such issues in the future. Every meeting should also have a facilitator who can enforce time limits and ensure the meeting stays on track. Establishing these norms can greatly increase meeting efficacy.

I'm sure you've gleaned by now that I have a strong aversion to ineffective meetings. They are not only a waste of time but also a financial drain for the organization. Hence, it's crucial to ensure meetings are fruitful. Always have an agenda. Invite only the essential participants. This not only streamlines the meeting but also enhances its effectiveness.

CONVERGENT, SHARED COLLABORATION

All right, diving into the realm of Shared Collaboration. This isn't just about working together; it's about striving for a heightened level of teamwork. Effective collaboration is rooted in communication, and not just the kind where everyone agrees, but the sort that embraces healthy disagreements.

Let's talk about diversity. Diversity fuels innovation. When I mention diversity, it isn't limited to cultural or ethnic backgrounds. It encapsulates a wide array of differences, including ideologies, perspectives, and skillsets. A team with diverse members is like a multi-faceted gem; each facet reflects a unique viewpoint, shedding light on unseen angles. This ideational diversity is a crucible for creativity and innovative solutions.

However, while diversity brings in a plethora of ideas, converging these into actionable plans is crucial. A team isn't successful if it's just a marketplace of varied thoughts; there needs to be convergence. This doesn't mean shutting down different ideas but rather amalgamating them to arrive at collective decisions. The goal is not individualistic assertions but a collective nod of agreement.

But there's a caveat. Proper facilitation is essential. Without structured methods, team discussions can devolve into endless debates, where decisions by committee become the norm, and more often than not, they aren't efficient. Numerous techniques can guide a team towards convergent decisions. For instance, Dot-voting can help prioritize ideas. Planning poker and Affinity Estimating aid in estimation tasks. Brainstorming sessions can be structured to ensure every voice is heard, and ideas are evaluated. We'll venture deeper into these facilitation techniques later.

At the core of shared collaboration is openness. It's essential to foster an environment where team members feel confident in voicing their disagreements. A simple "I don't think this is the right way to go" should be seen as a valuable input, not dissent. Reaching this level of mutual respect and understanding might take time, but it's an investment worth making for the betterment of the team and the projects they undertake.

PARTICIPATORY DECISION MODELS

All right, **Participatory Decision Models**. Here are some techniques they just talked about. So, diving into the realm of Participatory Decision Models, let's explore some practical techniques that can streamline group decision-making. The challenge with collaborative decisions often lies in reaching a consensus without suppressing individual perspectives. Let's break down some effective approaches:

1. **Simple Voting**: This technique is straightforward - raise your hand to vote. While it's uncomplicated, it may sometimes cut short deeper conversations. Remember, it's just one tool among many.
2. **Thumbs up, thumbs down, and sideways**: This approach provides a more nuanced perspective. A thumbs up indicates agreement, thumbs down signifies disagreement, and a sideways thumb implies uncertainty. This method allows for a quick temperature check on whether a topic needs further discussion or if consensus is already reached.
3. **Fist of Five**: This is a personal favorite. By showing anywhere from zero (strong disagreement) to five fingers (strong agreement), participants can visually express their level of agreement. It's a rapid way to gauge collective sentiment. For those with a hand disability, there are adaptive methods to ensure inclusivity. If the collective score indicates low agreement, dive into a deeper discussion. High scores might still warrant checking in with the few dissenters before moving forward.
4. **Dot-voting**: An efficient tool when confronted with multiple options. Here's how it works: given a list of choices, each participant gets a set number of dots (usually 3 or 5). They can distribute these dots across the options or place all on a

single choice. The option garnering the most dots typically emerges as the winner, but the process can also illuminate areas for further discussion. This method is particularly useful for prioritization tasks.

Each of these techniques has its strengths and can be immensely valuable depending on the situation. While making decisions, it's essential to strike a balance between swift consensus and ensuring every voice is heard. These tools are designed to aid in achieving just that.

Alright, let me share an anecdote that illuminates the effectiveness of these decision-making techniques. Some years ago, I was involved with a team aiming to conduct a JAD (Joint Application Design) session. Their mission was to collaboratively map out the work they'd be focusing on for the upcoming year, encompassing aspects like task size, priority, and effort required. To achieve this, they'd invited a sizable group of about 30 people to convene in Charlotte, North Carolina. The commitment was evident, both in terms of time and resources: a week-long session that wasn't easy on the pocket.

Witnessing their approach, I was convinced there were more efficient paths to the same destination. I proposed a challenge: allow me just 20 minutes at the outset, and I believed we could cover the majority of what they intended to discuss throughout the week. Met initially with incredulous looks, I unveiled my strategy.

Step 1: **Brainstorming**. We dedicated 5 minutes to capture every conceivable task or objective for the year, jotting down ideas on post-it notes.

Step 2: **Affinity Mapping**. Over the next 2 minutes, we grouped these notes, layering them based on their similarity. This allowed us to cluster related concepts, filter out redundancies, and whittle down to a more concise list, all in 7 minutes.

Step 3: **Dot-voting**. I distributed five dots to each participant, directing them to allocate their votes based on what they perceived as most pressing. Within 5 minutes, we had a prioritized list of tasks for the year.

Step 4: **Affinity Estimation**. The remaining time, slightly over 5 minutes, was used to gauge the scope of each task. This wasn't a precise measurement but a comparative one: was Task A bigger, smaller, or equivalent to Task B? Soon, we had a visual representation of our year's work in terms of importance and scale.

Within the initial 20 minutes, we had mapped out the year's probable work trajectory, its significance, and anticipated effort. This might not have been the final picture, but it was a strong starting point. The subsequent discussions then revolved around refining these details, providing clarity, and eliminating irrelevant tasks. The result was a more lucid understanding of our path forward.

Such facilitation techniques aren't only reserved for Agile environments but can be universally applied to make meetings and decisions more effective and efficient. Keep these tools in mind the next time you're faced with a daunting committee discussion. They're designed to get you to your destination faster and with clarity.

QUESTION 29

An agile team has ______ team members.

A. Have a large number of people (typically 24 or less)

B. Have a small number of people (typically 12 or less)

C. Have a small number of people (typically 5 or less)

D. Have a large number of people on rotation (120 people or more)

I will point out the Scrum Guide says between 3 and 9, but then there's the reality of what actually happens in the real world, and you typically see teams go up to maybe 12 and you can get away with that. You're not breaking any major rules by doing that. Just consider the bigger the team, the more complexity. And once you're getting to that point, you might want to consider breaking the team up into 2 separate teams.

Answer: The correct answer is B, 12 or less.

QUESTION 30

Which of the following is true about accountability in a Scrum Team?

A. The Product owner holds the development team accountable

B. Team members hold themselves mutually accountable

C. Team members hold themselves individually accountable

D. The Scrum master holds the team accountable

The Product Owner doesn't hold the team accountable. They're manager, they have no accountability for the team so they really can't tell the team that they're not doing their job. Of course, they would have discussions, right? If the Product Owner is not getting what they want then they would have a good collaboration about that, but it's about the team holding themselves accountable, right? So the correct answer here is B, and this is really important.

We want the teams to hold themselves mutually accountable, which means it actually include C. So, there's a little bit of a trick question, right? The statement is mutually accountable, meaning we held ourselves and each other, that's the mutually accountable, right? Ourselves and each other accountable for the work. We're going to give the person or people that said C, we'll give them a little bit of a break, right? But the correct answer would be B, and this is really, really important.

If you want transparency, if you want honesty, the team has to hold itself accountable. Otherwise, you're going to have to put somebody over top of that team to make sure that they are being held accountable. This is kind of a Theory X, Theory Y discussion, right? You start with Theory Y saying the team wants to do a good job and they can hold themselves accountable to do that good job. You don't need someone over top of them beating them over the heads. They will do the job and trust them to do it. And the good part is you have all this visibility and transparency into what they're working on so if they go off the rails you will see it, right? As long as you get that transparency, you don't need to tell them what to do.

Answer: Team members hold themselves mutually accountable

QUESTION 31

In the 5 stages of team development where do most arguments and disagreements happen?

A. Forming

B. Storming

C. Performing
D. Norming

Conflict is inevitable, isn't it? Whenever two individuals come together, disagreements and differing viewpoints are bound to arise, even if they're closely acquainted and trust one another. This natural ebb and flow of opinions is part and parcel of human interactions.

In group dynamics, this process has distinct phases. When we start in the **forming** stage, there's a certain delicacy and caution in our interactions. Think of it as the initial dance where everyone's still finding their footing, trying to avoid missteps and inadvertently stepping on someone's toes. It's the phase where politeness is at its peak, and there's a hesitation to rock the boat.

However, as comfort levels increase, we transition to the **storming** phase. This is when true colors start to show. Individuals become more vocal about their perspectives, and some might try to establish dominance or assert their viewpoints more strongly. The atmosphere can get charged, leading to more intense disagreements, and yes, occasional confrontations.

But, when the team matures and progresses to the **performing** stage, the dynamics shift again. Here, disagreements still persist, but they're often more constructive. Instead of unbridled arguments, you have more of debates where the end goal is common growth. It's about finding the best path forward for everyone involved. Disagreements are welcomed, as they bring diverse perspectives to the table. The focus shifts from "I'm right" to "What's right?"

Remember, disagreements are essential. They stimulate thinking, encourage diversity of thought, and often lead to better, more informed decisions. The key is ensuring they're healthy, constructive, and geared towards common objectives.

Answer: Storming

QUESTION 32

Which of the following are characteristics of high performing teams?

A. They are not self-organizing, rather they are role or title-based
B. They are not empowered to make decisions
C. Believe that as a team they cannot solve any problem
D. They are motivated by trust, instead of fear or anger

This was a good trick question. So obviously our answer is that they're motivated by trust instead of fear or anger. The others are very negative things. They're not self-organized. We want staff organizing. They're not empowered obviously and they should be able to solve any problem or at least some solve some problems, right? So C is worded in a way to kind of confuse, which was the intent, right?

Answer: D. They are motivated by trust, instead of fear or anger

We had just got done talking about building teams and we often think about the teams as the people building the work, building the products. We're also saying a team is also leadership, so we want our leadership to work very well as a team as well, so they should go through the same process. Learning how to disagree with each other, learning to have healthy conversations, learning about holding each other and themselves accountable not just holding their teams accountable. This is really difficult to put leadership in this position. They often don't like that but it really needs to be encouraged. In fact, that vote, the 5 dysfunctions is not about the team building the work. It's about a leadership team and how to make them more effective as well.

We view teams as not just people building it but also teams of leaders, right? We call them typically leadership teams, but what do we do to actually encourage them to work as an actual team? Often, I work with a lot of companies where that is just not a thing. Everybody is their own, once they get up at the leadership change, it's very siloed and very politically correct, and we want to discourage that. They have to work together as a team and disagree with each other in a healthy way.

CHAPTER 10

ESTIMATING IN AGILE

ESTIMATING – IDEAL TIME

Let's talk about **Estimating.** All right, so now, the first thing is estimating. We estimate differently in Agile. First thing is you do not have to use this technique to consider yourself an Agile team. You can estimate like you've always estimated but I'd like you to get this understanding and maybe you will see that this actually could be much more effective, much more accurate as a result.

In Agile, the method of estimating is innovative. It's pivotal to understand that adopting this technique is not a prerequisite to being an Agile team. You can retain your existing estimation methods, but embracing Agile's technique could enhance accuracy and efficiency.

The central question is, why opt for a different method of estimating in Agile? In Agile, estimating by time is not a preference, as it poses substantial challenges. Let's explore this with a basketball game analogy.

Here, we have a basketball clock.

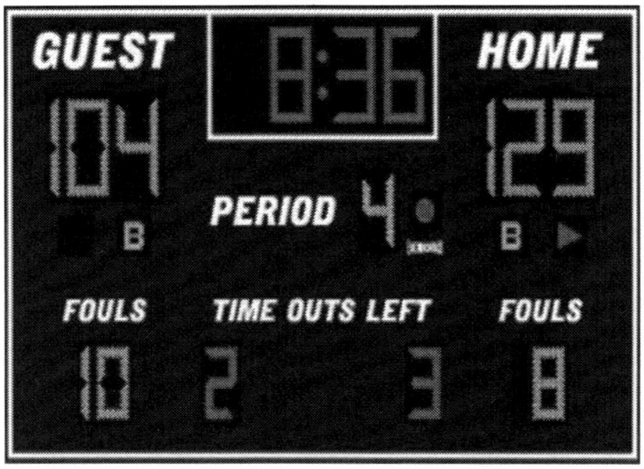

How much longer do you think this game is going to run? It says 8 minutes and 36 seconds. What are your thoughts? So the actual game time is 8 minutes and 36 seconds, but how much actual time is it going to take to complete this game? We've got fouls, we've got, you know, time out, so what do you think how long is this game going to run? Did I hear you say longer than 8 minutes and 36 seconds? Exactly right! the actual time to complete the game is going to vary from the ideal time!

The clock displays 8 minutes and 36 seconds, indicating the official game time left. However, considering fouls, timeouts, and other interruptions, the actual completion time invariably extends beyond the displayed time, highlighting the discrepancy between ideal and actual time.

This concept is mirrored in real-world scenarios. Work may be calculated to take a specific amount of "ideal" time, but the culmination of impediments, priorities, and life events invariably extend the "calendar" time required to complete the tasks.

For example, if I assume a task will take me three days, based on my past experiences, the requester might interpret it as three calendar days, creating unrealistic expectations. Given the myriad of projects and personal commitments, the completion might actually take much longer. To cushion ourselves against these discrepancies, we tend to pad our estimates, which often get manipulated further up the chain, leading to varied expectations and eventual accountability issues. This pattern induces stress and sets a tone of failure even before the commencement of a project, instigating a host of negative behaviors.

So, is there an alternative, more effective way to estimate? Enter **Story Points**. Story Points are pivotal in Agile and are not based on time but rather on the size and the complexity of the work. This method is used to create a relative estimate, gauging the size and complexity in relation to previously completed tasks.

Figure 25: The Concept of Triangulation

Consider a scenario of painting a room. If the first wall takes 2 hours, being a unit-based relative estimate, the similar-sized wall would also take 2 hours, and the walls that are twice the size would take 4 hours each. However, this method falls short if the baseline estimate is incorrect, leading to a re-estimation of the entire backlog.

To address this, Story Points are used, representing the "bigness" or the size and complexity of the work. A small wall can be represented as 1, not depicting any unit of time but rather the relative size. The larger walls, in this case, would be 2, maintaining the relative proportions. So, even if the time taken varies due to different factors, the relative sizes or the Story Points remain constant.

Adding a level of complexity, like a window, to the wall might increase the complexity while the size remains the same, reflecting the versatile application of Story Points in managing various facets of a task.

This approach, encompassing the relativity of size and complexity instead of a stringent time frame, can potentially lead to more accurate and adaptable estimating, facilitating enhanced project management and execution in Agile environments.

ESTIMATING – STORY POINTS

Now, in our example, instead of calling my estimate a 1, I'm going to call it a 2 because the complexity has increased, okay? That's relative estimating using Story Points. Now, here's the thing, we still need to get to time. You're never going to be able to go to your stakeholder and say "Yeah, that's about 50 Story Points, thanks" and then move on. They're going to want a date, right?

Now, if you ever heard that myth that Agile teams can't give you dates, don't believe it. We like dates. We are fixed time boxes. We like dates and we can give you very accurate dates if we're following good practices, okay? You can be much more predictable and accurate as you get better at this.

ESTIMATING – PLANNING POKER

All right. Let's move on. Let's talk about some techniques. The first one is called Planning poker. Planning Poker is a game-changing estimating technique used in Agile. It provides a platform for collective decision-making and helps ensure that team members are on the same page. Let's dive into how this works.

1. **What is Planning Poker?**
    - Planning Poker is a collaborative estimation technique. Rather than relying on just one individual's estimation, the entire team collaborates and provides their insights.
    - Every team member is provided with a set of cards, each displaying a number. These aren't just any numbers; they are taken from the Fibonacci sequence.
2. **Why Fibonacci?**
    - The Fibonacci sequence is a series of numbers where each number is the sum of the two preceding ones. It goes like this: 1, 2, 3, 5, 8, 13, and so on.
    - What makes Fibonacci special in this context? As the numbers grow, the gap between them widens. This reflects how our confidence level drops as tasks get bigger and more complex. A task estimated at 13 points is inherently more uncertain than one estimated at 2.
3. **How to Play?**
    - First, the Product Owner presents a task (or a "story") to the team.

- Team members ponder on this, thinking about its complexity relative to tasks they've done in the past.
- They then select a card that they feel represents the task's complexity.
- On a count, everyone reveals their cards simultaneously. This simultaneous reveal ensures that no one is unduly influenced by another's estimation.

4. **Interpreting the Results:**
   - If estimates are close, for instance, most cards are 2s and 3s, the team can simply agree on a number and move forward.
   - However, if there's a wide discrepancy (say, someone shows a 1 and another a 13), the team must discuss the reasons behind their estimations. Post-discussion, another round of voting can be done to see if a consensus is achieved.
   - If consensus remains elusive, it indicates the team needs more clarity on the task. It's better to pause and seek that clarity rather than forge ahead with uncertain estimates.
5. **Key Takeaways:**
   - Planning Poker promotes collective decision-making. Everyone's voice and perspective are taken into account.
   - The method focuses on one inclusive estimate for a task, not fragmenting it based on departments or roles.
   - It's not just about numbers; it's about understanding, collaboration, and making informed decisions as a team.

Planning Poker is more than just a game of cards. It's a strategic tool that fosters team collaboration and ensures everyone is aligned in their understanding and expectations.

And that's Planning poker! It's a simple process. You get to convergence on the estimate. It's a whole team estimate 1 number for all of the work to get done not multiple numbers. It's not an estimate for the development, not an estimate for verification. It's 1 number to consider the work done.

AFFINITY ESTIMATING

Let's explore another compelling estimating technique known as Affinity Estimating. I favor this method due to its swift and efficient nature, allowing teams to conceptualize scope and effort rapidly. We're expeditiously covering this as it's pivotal for our impending exercise.

Overview:

Affinity Estimating eschews numerical values, mitigating the common tendency to correlate '1' with '1 day'. The method encourages thinking in abstract terms—utilizing T-shirt sizes as metaphorical representations of task magnitudes. The abstract nature aids in deflecting from a numbers-focused mindset to an emphasis on relative sizing—Extra Small, Small, Medium, and Large. This facilitates a more intuitive, less restrictive understanding of task size.

Process:

1. **Seeding the Board:** The initial phase involves placing previously worked items, agreed upon as mediums, extra smalls, etc., on the board, serving as references.
2. **Sorting New Work:** Team members assess new tasks against the reference ones, conducting a comparative analysis to position the new tasks appropriately.

3. **Evaluation and Discussion:** Subsequent to task placement, the team collectively scrutinizes the board, raising challenges and instigating discussions regarding discrepancies in task placement.
4. **Product Owner's Challenge:** Though not involved in the execution, the Product Owner can raise discrepancies and trigger discussions if there's perceived incongruity in task sizing.

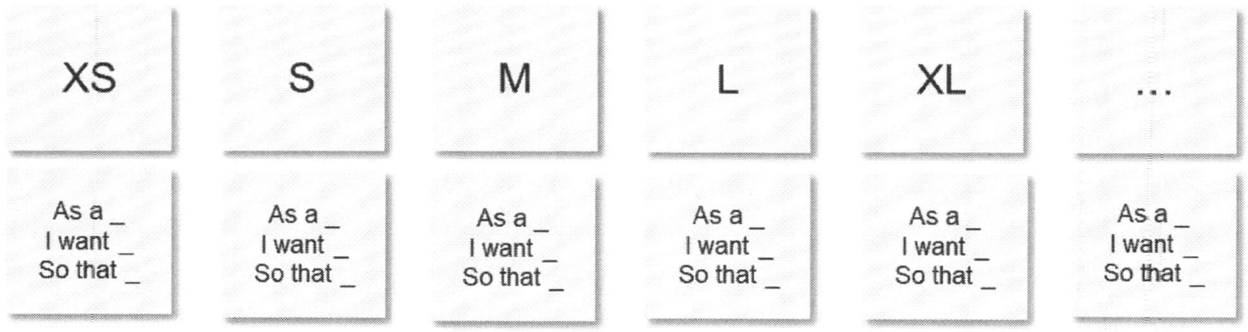

Figure 26: Affinity Estimating Example

Key Features:

- **Abstract Thinking:** Utilizing T-shirt sizes promotes abstract contemplation, avoiding direct correlations to days.
- **Dynamic Conversations:** The method encourages interactive discussions, resolving differences, and fostering mutual understanding.
- **Speed and Efficiency:** Affinity Estimating stands out for its quickness and ease, making it a preferred choice for teams aiming for rapid conclusions.

Conversion to Fibonacci:

Post Affinity Estimating, for effective forecasting, T-shirt sizes are aligned with Fibonacci numbers; for instance, an Extra Small could correspond to a 1, Small to a 3, Medium to a

5, and Large to an 8. This conversion is paramount for subsequent calculations and projections in the project's lifecycle.

Affinity Estimating is a versatile, fast-paced estimating approach, focusing on relational comparison rather than numerical fixation. It enhances dialogues, harmonizes perspectives, and swiftly converges to a consensus, positioning it as an invaluable tool in the agile estimating landscape.

VELOCITY

Let's talk a little about velocity and capacity. Let's talk about the difference. Velocity is the total amount of work that we complete by the end of the sprint. Now, we were less concerned about the work we didn't complete from a metrics perspective. At least from a velocity perspective, we're more concerned about what we did complete.

Velocity Overview:

- **Definition:** Velocity is the sum of the Story Points completed at the end of a sprint. It primarily focuses on what was accomplished, offering less concern from a metrics standpoint about uncompleted work. However, uncompleted work is crucial as it aids in gauging predictability.

- **Purpose:** Velocity is pivotal for forecasting. It's assessed across multiple sprints to discern trends and to stabilize it as a metric, thus aiding in improvement and ensuring predictability.
- **Misuse:** It shouldn't be used as a performance metric between teams as it can lead to gaming the system and provides no real comparative value due to variances in team dynamics, problems, experience, and estimates.

Capacity Overview:

- **Definition:** Capacity refers to the actual time available to execute work. It fluctuates each sprint due to variables like holidays, meetings, and training sessions affecting available work hours.
- **Purpose:** Capacity Planning is crucial, especially for new teams, to understand realistic availabilities and plan accordingly, rather than base planning on overly optimistic expectations.
- **Process:** It involves accounting for six work hours per day (excluding daily non-task-related activities) and subtracting hours spent on Scrum events, holidays, and other non-backlog-related tasks.

Interrelation between Velocity and Capacity:

- **Impact:** Capacity invariably impacts velocity. The varying available hours in different sprints will affect the amount of work completed, thus affecting the velocity.
- **Consideration:** Both aspects are vital to consider, and they must align to reality rather than optimism, ensuring teams are neither overcommitting nor underdelivering.

Capacity Planning and Reality:

- **Realization:** Teams often realize they have less actual time available than anticipated. Using physical representations like poker chips can visualize this disparity.

- **Initial Steps:** Early adherence to Capacity Planning enlightens teams to reality, allowing a smoother transition to using velocity for planning in the later stages.

Estimation and Commitment:

- **Tasks and Stories:** While stories are estimated in Story Points without knowing the executor, tasks are estimated in ideal hours, considering the volunteer for the task.
- **Commitment Limit:** When the available capacity approaches zero, teams should cease to commit to new tasks, keeping estimations intact and avoiding overcommitments.

Stakeholder Interaction and Date Predictions:

- **Date Forecasts:** While stakeholders often desire date predictions, Agile teams can and should provide date estimates using their velocity metrics.
- **Predictability:** By giving dates based on velocity, Agile teams can maintain predictability and credibility.

In Agile frameworks, understanding and leveraging velocity and capacity are crucial for optimizing productivity and ensuring realistic and achievable commitments. These metrics, when correctly interpreted and implemented, offer a balanced and transparent approach to sprint planning and execution. Additionally, they aid in eliminating detrimental practices like overcommitment and gaming the system, fostering a more harmonious and productive environment focused on continuous improvement and delivery of value.

FORECASTING TIME AND SCOPE

Agile forecasting is a method that utilizes empirical and historical data to predict project timelines and deliverables, aiding in effective decision-making. Below is a comprehensive explanation, using a chart as an example, depicting how forecasting can be achieved, focusing on sprints and story points.

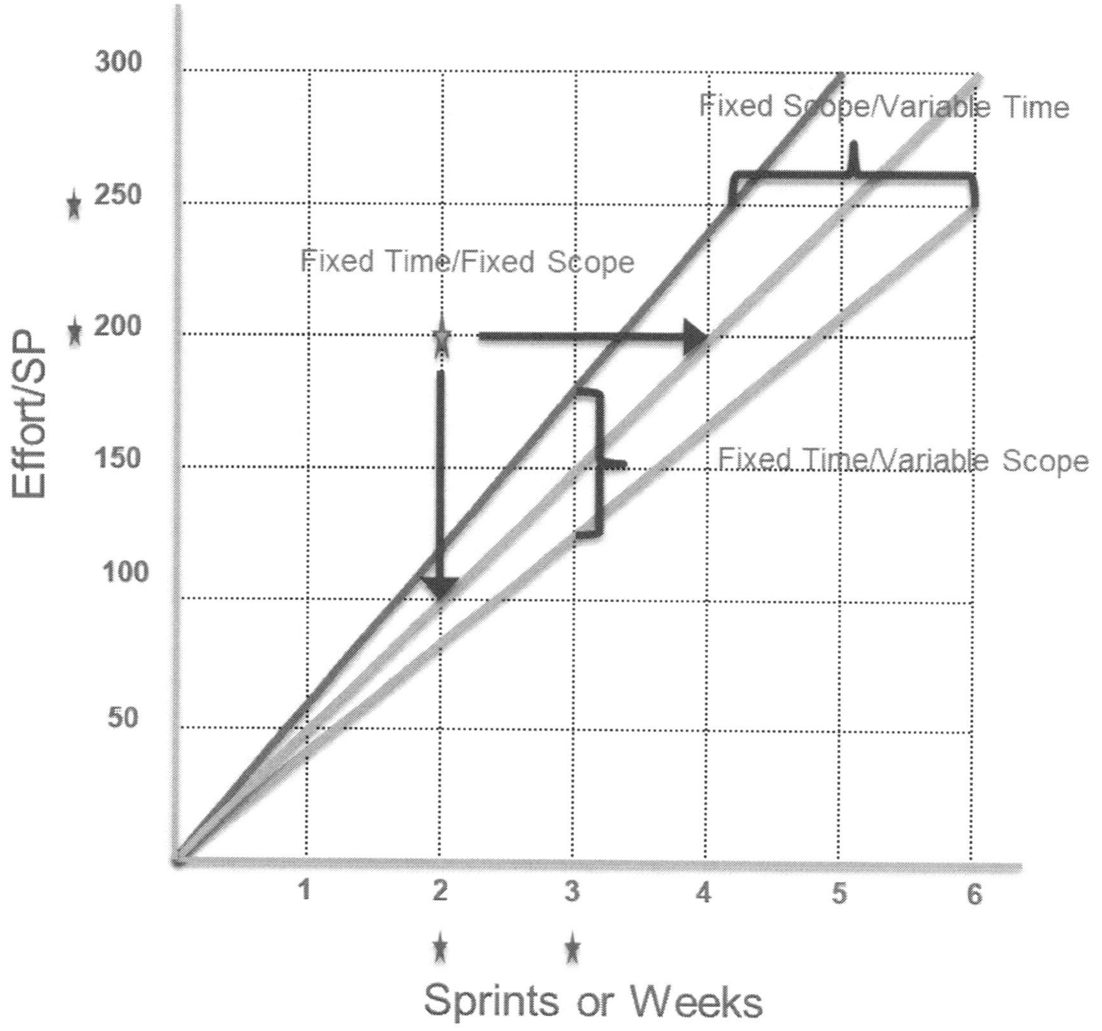

Figure 27: Forecasting & Velocity —in Agile

The previous figure illustrates the concept of Forecasting & Velocity in Agile. In this chart, the bottom axis represents time, measured in sprints or weeks, with each number signifying a sprint, equivalent to two weeks. The left axis denotes effort, expressed through story points.

Setting Up the Chart:

To commence, draw the average velocity line, representing the team's average points per sprint. This line won't be perfectly linear as it's expected that the team's velocity will incrementally increase over time if they are provided with adequate support, thus maintaining a slight upward slope. This helps in formulating more accurate forecasts, emphasizing consistent and sustained pace. The chart also includes optimistic and pessimistic lines, representing the best and worst velocities over time, respectively.

Cone of Uncertainty:

This concept elucidates that the further out the forecast, the higher the uncertainty and variability. A narrow gap between completed and committed points equates to higher predictability, while a large gap signifies lower predictability, leading to a wider cone of uncertainty. Maintaining consistency and reliability within teams is crucial to avoid substantial fluctuations in velocity, enhancing forecasting accuracy.

Using sprints as a time frame, each number at the bottom of our chart represents a sprint, with each sprint being two weeks. For example, sprint one is 2 weeks, sprint two is 4 weeks, and so on. The bottom of the chart represents time (sprints), and the side represents effort, which, in this context, is measured in Story Points.

To begin setting up the chart, we first draw our average velocity, representing how many story points a team can complete per sprint. Over time, a team's velocity might slightly increase, but this line isn't perfectly straight. If the team is well-supported and has the necessary resources, the velocity should show a slight upward trend over time.

From this average, which for the sake of example could be 50 points per sprint, you can establish an optimistic and pessimistic line. The optimistic line represents the best velocity over a given period, while the pessimistic line represents the worst. But, when considering the worst velocities, one should discard outliers like the holiday season, which is an anomaly.

This establishes a range of forecasting - the optimistic and pessimistic lines give you a fast and slow estimate. This range embodies the cone of uncertainty: the further out you forecast, the less precise your predictions will be. Your predictability is affected by the difference between committed points and completed points in a sprint. A large disparity indicates low predictability and a wider cone of uncertainty. Ideally, teams should maintain a consistent and reliable pace, avoiding wide fluctuations in their velocity and sticking to realistic commitments.

Considering scenarios can help illustrate this concept:

1. **Scenario - Fixed Scope, Variable Time**: If a stakeholder provides a specific amount of work and asks when it will be completed, the trend lines can provide an answer. For instance, the work might be completed between sprints 4 and 6, with sprint 5 being the most likely.
2. **Scenario - Fixed Time, Variable Scope**: When stakeholders want to know how much can be done by a certain date, the trend lines can provide a scope estimate.

For example, by a specific sprint, the team might complete between 125 to 175 points, with an average of 150.

3. **Scenario - Fixed Time, Fixed Scope**: If a stakeholder demands a certain scope by a particular date, the trend lines can reveal if it's feasible. If the desired scope surpasses the team's historic output, it's unlikely the goal will be met.

It's crucial not to outright reject stakeholder requests. Instead, suggest scope reduction, time extension, or both. Often, the instinctive solution is to increase team size. However, increasing team size can lead to reduced velocity and disrupted team dynamics, at least temporarily.

Rather than haphazardly adding team members, consider tapping into the capacity of existing teams. Maybe another team, with the right skill sets, can take on some of the work. If more capacity is genuinely needed, consider forming new teams. But it's important to remember that new teams take time to gel and become effective.

Forecasting in Agile, as described, uses actual historical data, making it a powerful tool for stakeholder communication. It visually represents what's achievable, providing a data-driven approach to predict and manage project outcomes.

QUESTION 33

Velocity represents _________

A. How much work the team can plan in one project management life-cycle
B. How much work the team can finish in one sprint
C. How much work the team can commit to in one project backlog
D. How much work the team can finish in one 24-hour daily scrum

The best answer to the question is option B. This defines velocity. How much work can you get done within a time box of an iteration Or Sprint.

Answer: How much work the team can finish in one sprint

QUESTION 34

In capacity planning, we plan for _________

A. 6 hours a day

B. 8 hours a day

C. 12 hours a day

D. 2 hours a day

Keep in mind, we're not saying that people don't work an 8-hour day. We're just saying that only 6 hours on average are available to actually do the work that's on our backlog. The other 2 hours and again, just the statistical average, it's just what most organizations are comfortable with. You're spending probably 2 hours a day just doing ancillary stuff, administrative overhead as part of your job. Just, you know, going for a bathroom break takes time, getting coffee takes time and answering emails. All that stuff takes time away from doing the work on your backlog and that's why we only factor 6 hours, 8 hours you're still working, 6 hours is what we plan following for.

Answer: 6 hours a day

QUESTION 35

In capacity planning, we do all of the following EXCEPT

A. Deduct Scrum event time

B. Deduct time off

C. Deduct time to take your dog to the vet

D. Deduct actual time available to work

Wow, it's one of those trick questions, well done. That's our goal, is to actually come up with that number, that's what we're trying to achieve. All those other things, take away from our available time, and obviously taking your dog to the vet is just an example but anything that takes you away from doing the work on your backlog. We're going to be realistic about it and remove it from your available capacity because it's reality. Funny fellow, I was once called a hope killer and when I first was called that, you know, I didn't know how to take it but I realized it's actually a compliment, right? Because we're not just planning based on hope. We plan on reality, the hope away and let's go with the reality plan.

I think it's also a challenge in the way that we estimate using days and weeks and things like that in predictive because the days and weeks that we're using really are not in any way related to the actual amount of time to do the work. It's related to a delivery date, and there's so many factors that go into a delivery date, other priorities and things that get in your way. There's so many variables to get you there that. Estimating in time causes all kinds of challenges with that.

Answer: Deduct time off

QUESTION 36

Which of the following is TRUE?

A. Stories are estimated in hours

B. Tasks are estimated in story points

C. Stories are estimated in sprint days

D. Tasks are estimated in hours

Think about it this way, the name or Story Points really tell you what they're used for, right? Story Points are used to estimate stories, right? Now, tasks are the steps that we're going to take in order to implement those stories. They're very, very small units of work. I generally try to keep my tasks down to around a day. We feel much more comfortable estimating those in hours simply because they're so very small. We know who's going to do the work and we have a pretty good idea, actually a very good idea of how we're going to get that work done. So stories and Story Points, tasks in hours.

Answer: Tasks are estimated in hours

QUESTION 37

Which of the following is true about velocity?

A. Different teams have different velocity
B. All teams have the same velocity
C. Teams in the same company have the same velocity
D. Never average the velocity across multiple sprints

Let's address this in reverse or let's talk about C, how healthy a team is. Velocity doesn't really indicate the health of a team. Certainly, it is the amount of output but what we're really looking for...one of the things we're looking for that indicate a healthy team is their improvement in their velocity, right? It's not just that, you know, if they're producing points, it's a healthy team, that doesn't really mean anything since it's not a performance metric. It's really are we getting better? We use a lot of this data to indicate are we getting better or not, and that really indicates more the health of a team.

In fact, something we don't really talk about in this class is doing team health assessments and things like that. We spent a lot of time determining the health of the various things in our organization, the health of your program, the health of a project, health of the product, health of the team, health of a lot of stuff, and we take survey and look at metrics of data to determine health to see if we're getting better. Velocity may be an attribute but it's by itself does not in any way indicate the health of the team, and just going back to something I said, velocity is also not a performance metric, it's a forecasting metric. The word performance might be a little confusing. Performance does not mean that they are how we're going to perform. It's instead how are we performing against other teams? That's really what we're looking at. Don't use it to measure teams. Don't use it to compare teams. That's what we're trying to get to absolutely.

Answer: D. Different teams have different velocity

KANBAN

Kanban, one of my favorite topics, is a straightforward process which originates from Toyota's lean production system, adapted for product development. While technically stemming from manufacturing, it's now recognized as an Agile process. For those keen on venturing deep into Kanban, David Anderson's book titled "Kanban" is a must-read. This foundational work, while not overly advanced, gives insight into its history and its varied applications.

Kanban translates roughly to "signboard." Central to the Kanban system is the idea of a continuous flow process. Differing from other models where work is assigned to teams, in Kanban, teams pull work into their process when they're ready, enhancing efficiency. A vital aspect of Kanban is limiting work in progress (WIP). This focus ensures faster task completion and emphasizes the flow of work from start to finish. By limiting WIP, teams can identify issues, minimize waste, and maintain focus. The aim? Get tasks done as swiftly as possible.

One of Kanban's strengths is its evolutionary nature. You begin with your existing process, roles, and methods. From there, a board is set up, and every task is represented by a card on this board, tracking its progress. No immediate drastic changes, just gradual evolution and improvement.

Kanban isn't limited to organizations; its principles can be applied personally. I use it for tasks ranging from course materials to household chores. With Kanban, you're essentially beginning with your current situation and making consistent improvements.

But as with all things, there are rules. Firstly, ensure work passed on is of the highest quality to avoid the trap of rework later. Each stage should focus on quality. Secondly, always pull only what's needed at the moment. This pull system ensures work moves at a steady pace. Before beginning any work, clear any possible hindrances to ensure uninterrupted flow. In essence, Kanban promotes a consistent flow of work and optimal productivity.

Kanban, inspired by lean manufacturing principles, has emerged as a potent remedy to Waterfall development's bottlenecks. The method's core principle of limiting "work-in-progress" ensures that teams aren't overwhelmed and can deliver products with efficacy and swiftness. By ensuring that no new task is initiated until the current one is complete, the system makes obstructions more noticeable, paving the way for more streamlined processes.

While Kanban's roots trace back to manufacturing, its principles have been successfully applied to software development, which is our focus here. A standout feature of Kanban is its absence of strict roles, granting teams the agility to interplay with each other's tasks and responsibilities effortlessly.

Central to the Kanban system is the Kanban board, a vital tool that gives a pictorial representation of work's flow. This board can be tangible, like a whiteboard with sticky notes, or digital, like many of today's project management tools. The board is divided into various stages of development, with each stage having a set limit on the number of tasks it can contain. This visual system, exemplified by sticky notes or cards, enables teams to move tasks from one stage to the next, providing an immediate overview of work progression and areas demanding attention.

At its heart, Kanban thrives on three foundational principles:

1. **Visualize Current Workflows:** By representing tasks visually, teams can better understand and manage their workload.
2. **Limit Work-in-Progress:** This ensures that teams aren't spread too thin, resulting in enhanced efficiency and output.
3. **Emphasize Flow:** By prioritizing tasks effectively, Kanban ensures a continuous and seamless flow of work.

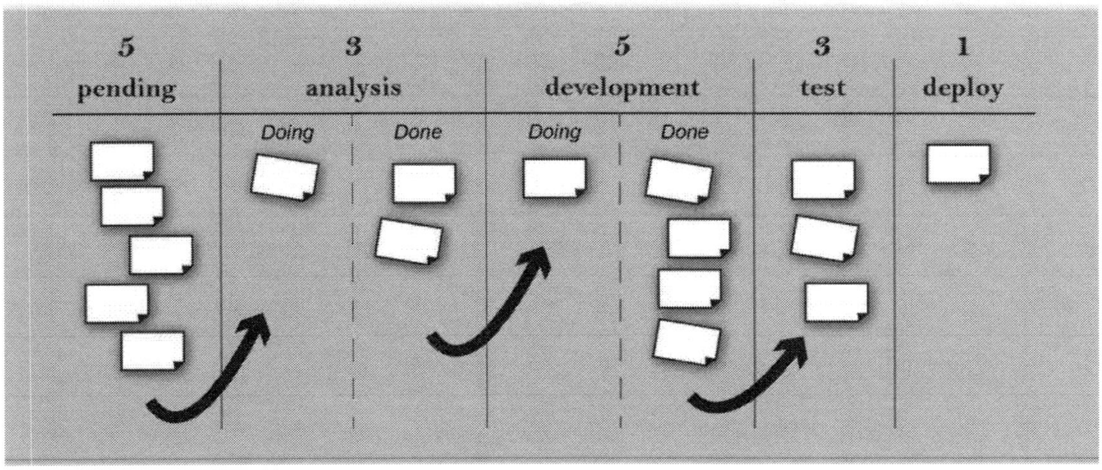

Figure 28: Example of a Kanban Board

One of the standout merits of Kanban is its adaptability to frequent changes, making it a preferred choice for projects with shifting requirements. Shorter cycle times facilitate regular customer feedback, enabling swifter alignment with user needs and faster feature delivery than conventional methods. Moreover, Kanban encourages uniformity in task sizes, streamlining work-in-progress and making it more controllable and foreseeable.

PRINCIPLES OF KANBAN

Venturing further into the core principles of Kanban can provide better clarity into its methodology and the secrets behind its success.

1. **Visualize the Workflow**
    - At the crux of Kanban is the visualization of work. Using a board to depict various states of the workflow, teams can quickly gauge the status of various tasks. The Kanban board becomes a central focal point, allowing team members to drag and drop tasks (often represented by cards) from one state to the next, providing real-time updates.
2. **Limit Work-in-Progress (WIP)**
    - This principle emphasizes the importance of controlling the number of tasks being executed concurrently. Surprisingly, engaging in more tasks simultaneously can delay the completion of each one. A pivotal reflection here is: *Is the goal to initiate numerous tasks or to finish them?* The answer underpins the essence of Kanban. Initiating numerous tasks may seem productive on the surface, but true value is derived when tasks are concluded. Thus, it's more beneficial to commence fewer tasks and see them through to completion swiftly. This might necessitate a reorientation in the organizational culture and mindset, shifting the focus from mere activity to productive completion.
3. **Manage the Flow**
    - Beyond just visualizing tasks, it's crucial to ensure a steady and efficient flow of work across the board. Effective management ensures that tasks

progress smoothly from one stage to the next, minimizing stalls and ensuring timely delivery.

4. **Make Policies Explicit**
    - Clarity is paramount in Kanban. Every transition of a task from one stage to the next should be backed by clear policies. These policies, or entrance criteria, dictate the conditions that must be fulfilled for a task to progress to the subsequent stage. Whether it's moving a task from 'To Do' to 'In Progress' or from 'In Progress' to 'Done', each transition should be defined clearly. This clarity ensures uniformity in understanding and prevents ambiguity, fostering a more streamlined workflow.
5. **Improve Collaboratively**
    - Kanban isn't a static method but a dynamic one. Regular introspections and team discussions should be encouraged to identify areas of improvement. The collaborative nature ensures diverse perspectives are considered, leading to holistic betterment.

These principles form the backbone of Kanban, guiding teams to achieve efficiency, clarity, and continuous improvement in their projects.

At its core, Kanban embodies the spirit of continuous improvement, a concept inherited from its lean manufacturing roots, famously captured by the term "Kaizen."

1. **Kaizen in Kanban**
    - Derived from the Japanese words 'Kai,' meaning 'change,' and 'Zen,' meaning 'good,' Kaizen stands for continuous improvement. While Agile

methodologies adopted this concept and termed it 'retrospectives,' the heart of the idea remains unchanged.

- Kaizen isn't solely about group efforts to refine processes. It strongly emphasizes individual improvement, fostering a culture where every team member strives for excellence, always aiming to better their work, methodologies, and output.
- Therefore, when practicing Kanban, periodic Kaizen events are crucial. These are dedicated sessions where the team convenes to discuss how they can enhance their work processes, efficiency, and collaboration. It's a cycle of reflection and refinement.

2. **Deep Dive into Kanban Workflow**
   - Much like other Agile methodologies, Kanban relies on a list of pending tasks, akin to Scrum's Product Backlog. Although not officially termed 'Product Backlog' in Kanban, the function it serves is identical. It's essentially a compilation of tasks, features, and jobs that await commencement.
   - The Kanban board is central to the methodology. Think of it as a visual representation of the workflow, where tasks transition from one stage to another. Every stage represents a specific phase in the project lifecycle.
   - To explain with an example: Let's visualize a Kanban board that starts with 'Queue,' signifying tasks that are in line for execution. The next phase could be 'Build,' where tasks are actively being worked upon. This could be followed by 'Test or Verify' where they undergo quality checks, then 'Review' where stakeholders or team members evaluate them, culminating in the final 'Done' stage.
   - Remember, the stages (columns) on a Kanban board aren't set in stone. They should accurately mirror the team's workflow. Therefore, while

beginners might find solace in a basic board with columns like 'To Do,' 'In Progress,' and 'Done,' as they become more attuned to Kanban, they can evolve their board to better represent their workflow.

- The beauty of Kanban lies in its flexibility and adaptability. It encourages teams to start simple, understand their work patterns, and then progressively refine their board and processes to maximize efficiency. The key takeaway is to adapt and evolve, ensuring the methodology serves the team rather than the other way around.

Kanban's synergy of visual workflow representation with the Kaizen philosophy ensures that teams not only get tasks done but are consistently refining their processes, optimizing for efficiency, and ensuring they deliver the highest quality results.

Work Decay in Kanban:

When work sits stagnant for a significant duration, it runs the risk of becoming outdated or irrelevant. This is known as work "decay". Factors like changes in the external environment, client needs, or project requirements can alter the relevance of a task that was once deemed crucial. So, the longer a task remains unattended, the more likely it is to lose its importance or require significant revisions when it's eventually addressed.

Addressing Overloaded States:

When states like 'build' and 'test' are overloaded and reaching their WIP limits, it's a clear indication of a bottleneck. Addressing bottlenecks in a timely manner is crucial in Kanban. The first step would be to pinpoint the cause. Are there not enough resources in the testing phase? Is the build phase producing work faster than the testing phase can handle? Answering these questions will help decide the next steps.

1. **Collaboration**: Team members from the 'build' phase could potentially collaborate with those in the 'test' phase to expedite work and help clear the backlog. This cross-functional approach ensures smoother flow.
2. **Re-evaluating WIP Limits**: Maybe the initial WIP limits were not set appropriately, and there's actually more capacity than initially assumed. If so, teams might decide to adjust these limits. However, caution should be taken to ensure quality isn't compromised.
3. **Pause and Reflect**: If tasks keep piling up and WIP limits are consistently hit, it might be worth pausing to hold a reflection or kaizen session. The team could discuss ways to streamline the process or even consider bringing in additional resources.
4. **Prioritization**: In cases where the backlog is extensive, it becomes crucial to prioritize tasks. By focusing on high-priority tasks first, teams ensure that the most valuable work is getting done, even when capacity is limited.

Maintaining the Pull System:

It's vital not to compromise the pull system. Pushing work into a phase that's already at capacity can be detrimental. It can lead to increased stress, a higher likelihood of errors, and decreased morale.

Wrap Up:

Kanban's flexibility and visual nature can provide teams with clear indicators when things aren't flowing smoothly. By respecting WIP limits, focusing on the pull system, and constantly aiming for improvement through kaizen, teams can navigate challenges effectively. It's all about balance: balance in workload, balance in capacity, and balance in delivering value efficiently.

In many project environments, there's a tendency for work to accumulate without reaching completion. This can result in 'inventory decay' or functionality built atop partially done tasks. The danger lies in the potential for requirements to become outdated due to new insights. This ultimately hampers the movement of tasks to a 'done' status, leading to a pattern of continually starting new tasks rather than completing current ones.

Rather than this unsustainable pattern, the Kanban approach advocates looking "to the right". For instance, if builders are at capacity, they should turn to testers to ask, "How can we assist you?" By addressing bottlenecks down the process line, we can ensure smoother workflow progression.

Even if builders lack the expertise to test, they can aid testers in other ways: setting up equipment, ensuring conducive environments, or removing blockers. The aim is mutual assistance within the team to expedite task completion.

Kanban's mantra is simple: "Stop starting, start finishing." Only when current tasks move to 'done' can new ones be undertaken. This way, there's a consistent flow of work. A crucial Kanban metric is Cycle Time, which measures the average time taken from a task's initiation to its completion. This data is invaluable for forecasting. Once a team has enough empirical data from using Kanban, estimation is replaced by actual Cycle Time data, making the process more reliable.

However, to harness the full power of Cycle Time, it's essential to recognize patterns in the nature of the tasks. Different tasks, or 'classes of service', have varied cycle times. Recognizing these patterns and categorizing them appropriately results in more realistic forecasting.

In addition to Cycle Time, there's Lead Time, which measures the duration from when a task enters the backlog until it's delivered to the customer. Imagine shopping on Amazon: the Lead Time is from when you order a product to its arrival. However, to ensure that Lead Time remains realistic, it's vital to maintain a clean backlog, regularly purging outdated or irrelevant tasks.

In essence, Kanban promotes a continuous flow process, making it versatile for varied industries, including those dealing with physical products. It offers the flexibility to re-prioritize tasks, as opposed to being constrained within fixed periods like sprints.

Consider this real-world scenario involving a software development team. This team was developing software for a hardware manufacturing company to enhance their production process. Their task flow encompassed several stages, from 'to do' to 'in production'—their desired end state for tracking tasks, which aligns with the fundamental approach of Kanban.

Despite doing an exceptional job in producing work efficiently, the team encountered a significant bottleneck in the "Ready for UAT" (User Acceptance Test) stage. While they consistently moved tasks to this stage, the separate UAT team, responsible for handling these tasks, couldn't keep pace. This disparity resulted in a mounting backlog at the UAT stage.

This situation harkens back to the Theory of Constraints. Imagine a manufacturing line with multiple workstations, where one particular workstation, situated in the middle, is the slowest. According to this theory, the pace of the entire process is determined by this slowest workstation. Any optimization before or after this point won't genuinely enhance the overall flow. In this software team's scenario, UAT was that limiting factor. Tasks that took the primary team about a week to complete took the UAT team a month. The true inefficiency wasn't with the software team, but the UAT process.

An initial proposition was to simply remove the UAT stage from their board. However, this wouldn't solve the problem; it would merely mask it. Even if the primary team seemed efficient, the delay in reaching the 'in production' stage remained.

To address this, they opted for a more strategic approach. Recognizing that it took an average of one week to move tasks from 'to do' to 'Ready for UAT,' they decided to sync with the UAT team in advance. Before initiating a task, they would communicate with the UAT team, estimating a week-long completion period, and inquire if the UAT team would be ready to test upon completion. If the UAT team wasn't prepared, the primary team would hold off, preventing unnecessary backlog accumulation. This proactive communication ensured better alignment and smoother workflow across both teams.

Our primary objective isn't merely to keep everyone occupied. Rather, it's to see tasks through to completion efficiently. If there's additional time available, instead of unnecessarily crowding the workflow, it might be beneficial to address technical debt, improve documentation, or assist the UAT team in enhancing their efficiency.

By adopting this approach, we not only minimized the backlog of tasks but also concentrated on streamlining the UAT process. This emphasis shifted from just improving the turnaround time for individual tasks to improving the overall duration it takes for tasks to progress from 'to do' to 'done.'

This holistic view of the process is where Kanban excels. It encourages teams to scrutinize the entire system, rather than only focusing on isolated segments. That wraps up the main points about Kanban.

QUESTION 38

Cycle time in Kanban is average amount of time to get work done

A. True

B. False

It is true, Cycle Time in Kanban is the average amount of time to get work done. Think about it this way, Cycle Time also needs to be divided up by class of service. It's not just one broad stroke average because different kinds of work are going to have different averages, but it's our primary metric that we use for forecasting and improvement in our Kanban process.

Answer A: True

QUESTION 39

Kanban is a _______ word meaning __________

A. Japanese, Signpost

B. Japanese, Signature

C. Chinese, Signal

D. Japanese, Center

Answer: A. Signpost

QUESTION 40

In Kanban, one of the five principles is __________

A. Maximize the WIP

B. Keep process policies loose

C. Improve independently – each person experiments individually

D. Minimize the WIP

Yeah, so it sounds counterintuitive, right? The concept that we typically work with as an organization is to do a whole bunch of things at the same time and it gives an illusion that we're getting a lot of work done, but the key word is done, right? We're starting a whole bunch of things, we're working on a whole bunch of things, and we're getting things moving, we're keeping busy, but it doesn't mean we're actually getting things done. So if you work on fewer things, and we're going to show this on the next slide, mathematically there's queuing theory and all kinds of stuff that goes to prove that working on the few things as possible will get work done faster.

Answer: D. Minimize the WIP

QUESTION 41

If wait time is too high in Kanban, it could indicate:

A. Bottlenecks

B. Value

C. High performance

D. High velocity

Answer: A. Bottlenecks

Wait-time typically means that things are waiting to get to done, so that means that we've got a bottleneck somewhere.

QUESTION 42

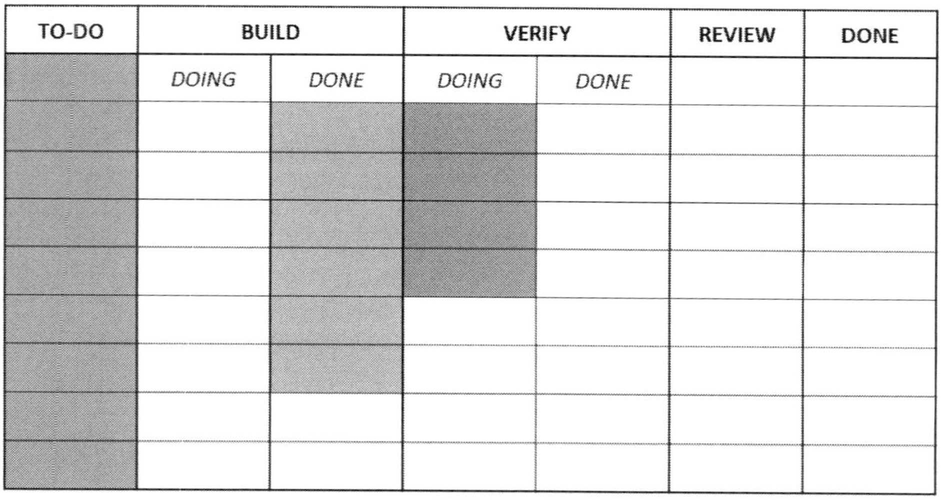

In the table shown above, build state WIP is 6 and is reached.

Verify state WIP is 4 and is reached

What should the team do?

A. Go back to the To-Do section and get as much work as possible

B. Focus on starting a chunk of new work. Ignore finishing for now

C. Look to the right to see what the bottleneck is and help accordingly

D. Look to the left always. New items can be pulled from To-Do to keep work in motion (WIP)

Most organizations think if you have run out of work go start something new, right? Go bring more work in because that's good. That means we're getting stuff done but it's exactly opposite of getting stuff done. All it means is that we're accumulating more work in progress, accumulating things just sitting there in inventory, waiting to be completed. The thing that we really want to do is look to the right and this is C, look to the right and see what we can do to remove that bottleneck. Help the rest of the team out so that we can remove the bottleneck and clear the flow of work so we have nice consistent flow to get done as fast as possible.

Answer: Look to the right to see what the bottleneck is and help accordingly

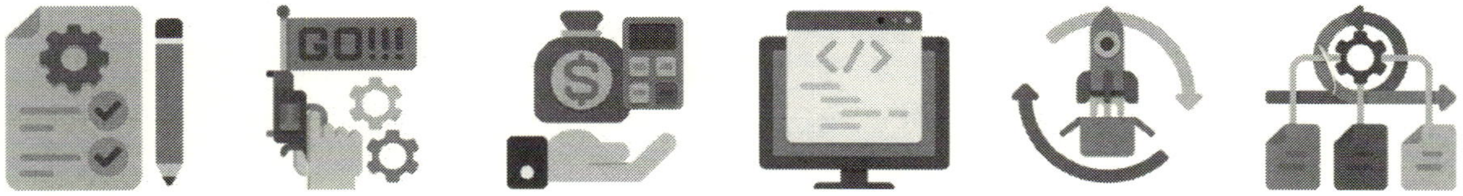

LITTLE'S LAW

Little's Law isn't termed a 'law' lightly; it's grounded in proven principles. Simplified, the law posits: the higher the work-in-progress (WIP), the slower the output and longer the completion time. You can watch our extensive YouTube video about Little's Law: (https://www.youtube.com/watch?v=4XamhNqP-7w), The basic concept reinforces the idea of reducing your WIP. This is some of the math behind this. Little's Law, is called a law for a reason not Little's theory, right? Little's Law basically states (paraphrasing), *the more work you have in progress, the slower you are and the longer it takes to get work done.*

It's pivotal to comprehend this: to enhance throughput and increase production, you must curtail your WIP. The math supports this. If you aim to expedite your process and produce more efficiently, the answer remains the same - restrict your WIP. This principle counters traditional beliefs, where emphasis is often on individual productivity and keeping everyone occupied. But it's crucial to realize that our endgame isn't about mere activity; it's about results.

Rather than initiating multiple projects or tasks, it's wiser to limit WIP. Adopting a Kanban perspective, focus on completion rather than mere initiation. Embracing this principle is fundamental to an organization's efficiency. By narrowing down your active projects and tasks, by really emphasizing finishing over starting, you'll witness substantial gains in efficiency. Remember, our core objective is completion, not just commencement. The true measure of success isn't in how many tasks are started but in how many are effectively finished.

I can't emphasize how core and critical this is to efficiency of an organization. Focus on limiting your work in progress, reduce the number of projects you're working on, reduce the number of things that you're doing together as a team, get it smaller, reduce the amount of work in progress, and you will be amazed at the efficiency gains.

You will get things done faster. "Done" is important, starting is not, okay? That's our goal. Many start but few finish!

COST OF DELAY

Cost of Delay (CoD) is an essential tool in project prioritization. At its core, CoD is an economic prioritization methodology that integrates both urgency (the speed we need things done) and value (the importance of the deliverable).

Understanding CoD:

Although termed a 'calculation', CoD heavily relies on subjective evaluations. The math might seem firm, but the numbers are inherently fuzzy. At its base, the CoD is calculated using factors like business value, time criticality, risk reduction, and opportunity enablement. The aim is to ascertain the duration we can postpone a task before it poses issues.

Components of CoD:

1. **Business Value:** Not directly proportional to monetary values but rather a gauge of relative importance when compared with other backlog items.
2. **Time Criticality:** This relates to non-negotiable deadlines, such as regulatory, legal, governance requirements, or even contractual obligations.
3. **Risk Reduction:** This speaks to the lowering of market risks or other similar threats.
4. **Opportunity Enablement:** This could be pioneering a new product or venturing into fresh market arenas.

CoD in Action:

Upon summing the above attributes, you get a CoD value. But, a cautionary note - these figures don't signify precision; they're guiding posts. One person's perceived urgency or value might differ from another's.

Extension - Weighted Shortest Job First (WSJF):

WSJF is another pivotal concept here. Simply put, I call it "the biggest bang for the buck." It's the CoD divided by the estimated effort (e.g., Story Points). Its essence s to prioritize smaller tasks with higher value. A small effort task that gives tremendous value comparatively, should be prioritized accordingly.

Understanding WSJF:

The idea behind WSJF is that tasks with a short duration but high value should be given priority. However, while WSJF provides a structure to backlog prioritization, it shouldn't be seen as an iron-clad rule. If taken too literally, only small, high-value tasks would be addressed, potentially sidelining other significant initiatives.

Conclusion: The beauty of CoD and WSJF is in their dual focus on urgency and value. Both serve as invaluable tools for structuring a backlog, ensuring teams are centered on what truly matters. But remember, they're guidance tools, not commandments.

VALUE AND URGENCY SIMPLIFIED

Consider the Value-Urgency Matrix as a simplified visual alternative to the Cost of Delay. The horizontal axis represents 'Urgency', while the vertical axis stands for 'Value'. It's like a quadrant system where items are plotted based on their relative urgency and value.

The idea here is similar to what we did with risk. You might say the things that are in the high range of value x urgency, perhaps those are the things that we want to work on first. But again, just a guide.

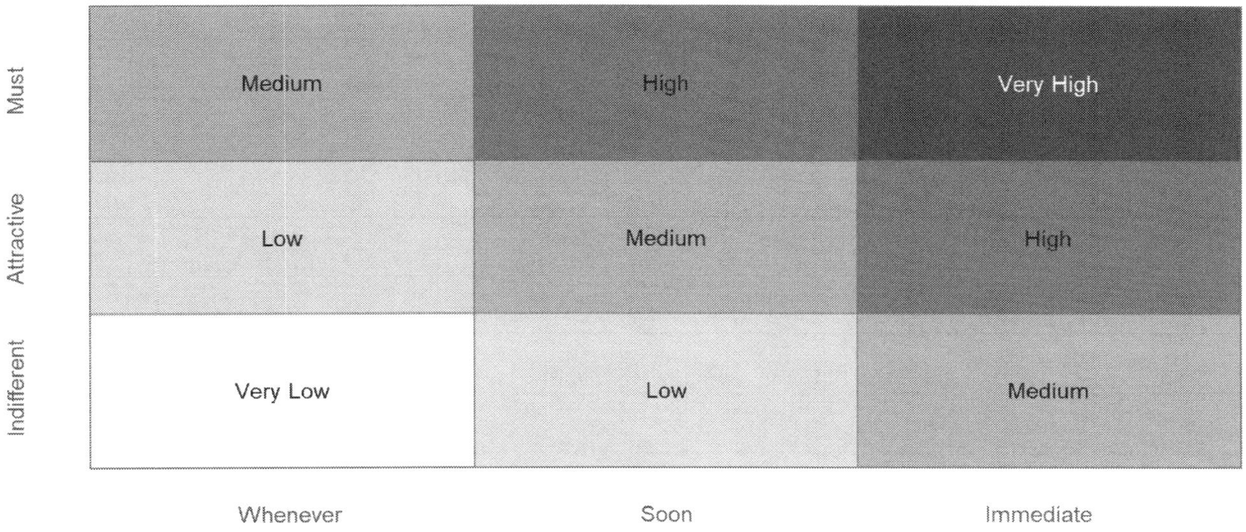

Figure 29: Value and Urgency Simplified

How to Use the Matrix:

1. **High Value & High Urgency:** Features or tasks that fall in this quadrant are crucial. They offer substantial benefits and need immediate attention. These are typically the top priorities.
2. **High Value & Low Urgency:** These are important items but don't have an immediate deadline. They can be scheduled after the high urgency ones but shouldn't be ignored for too long.

3. **Low Value & High Urgency:** Tasks here might not have significant long-term benefits but need to be tackled immediately due to external constraints or short-term gains.
4. **Low Value & Low Urgency:** These tasks are of the least priority. They can be scheduled when resources are available or deferred for a later time.

Why the Matrix Works:

The beauty of this matrix is its visual simplicity. At a quick glance, teams can gauge what needs attention and when. Decisions become more streamlined, and teams can often avoid the complexities of detailed calculations.

Conclusion:

The Value-Urgency Matrix provides a visual and straightforward approach to prioritizing tasks. While it might not provide the in-depth analysis that the Cost of Delay does, it offers a quick, intuitive method for teams to organize their backlog and identify what to focus on next.

FLAVORS OF AGILE

As mentioned at the beginning, there are tons of different practices out there, Scrum and Kanban are the two main ones, we talked about. I also want to highlight Scrumban. Scrumban is a combination of Scrum and Kanban as the name kind of implies. So let's explore this and a few more Agile approaches. Below is a pie chart of relative use of the main Agile frameworks and methods.

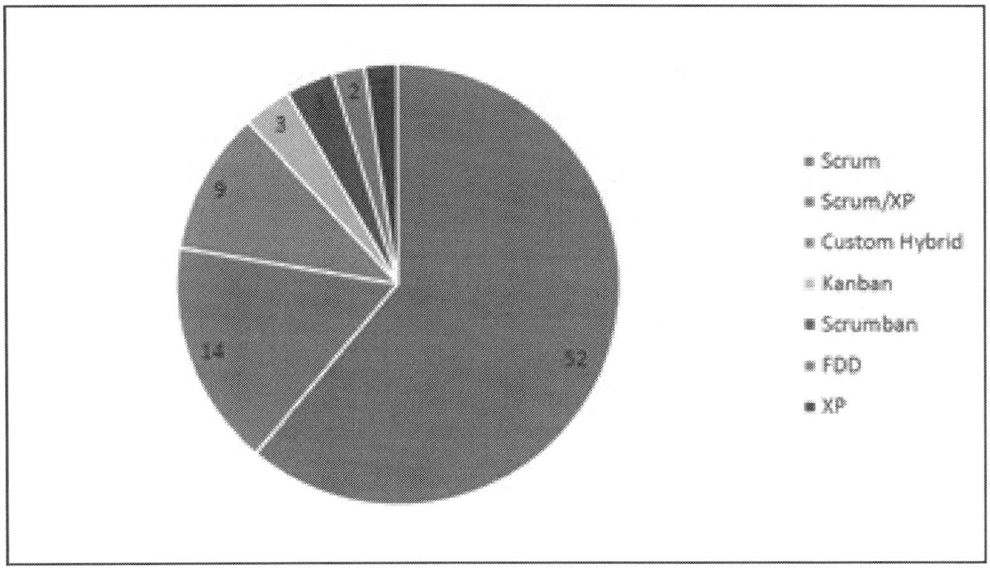

1. Scrumban:

Scrumban combines both Scrum and Kanban, typically by using the Scrum team structure with Kanban process principles. Scrumban is seen as being more flexible than Scrum, but more structured than Kanban. Similar to Scrum, Scrumban uses iterative planning, requirements prioritization, and structured teams. From Kanban, Scrumban uses the pull system, work-in-progress limits, and work visualization (Kanban board). Scrumban relies on the principles of Scrum and Kanban, as discussed in the previous sections. Scrumban is a Kanban practice. Meaning it is continuous flow, has the exact same rules as Kanban but we tailor up, we add some additional capabilities from Scrum so that we get some of the benefits that we get from Scrum. Scrum is focused mostly on project management. Kanban is focused on flow of work, getting things to done faster. The two together are really effective and there are other forms now. The industry is still adapting and changing, which we want to encourage. You should always adapt to change.

Emerging from a need for both structure and flow, Scrumban is a potent fusion.

- *Hybrid Approach:* Scrumban melds Scrum's iterative planning and team structure with Kanban's focus on flow and task completion.
- *Benefits:* More flexibility than Scrum, structured flow from Kanban.
- *Key Concepts:* Continuous flow, pull system, Daily Standups, Retrospectives. and Backlog Refinement sessions from Scrum.
- *Queue Replenishment:* Rather than Sprint Planning, work gets added to the queue as tasks are completed.

2. eXtreme Programming (XP):

With its roots in coding, XP ensures teams stay closely connected to the project's heart: the code.

- *Team Structure:* XP focuses on self-organized, cross-functional teams that include the customer.
- *Best Practices:* Small releases, simple design, pair programming, and continuous feedback loops.
- *Goal:* Reduce cost of changes in requirements and enhance code quality.

3. Dynamic Systems Development Method (DSDM):

As one of the earliest agile methodologies, DSDM underscores the importance of full life cycle management.

- *History:* Introduced in 1994, DSDM emphasizes transparency and communication.
- *Phases:* Feasibility, foundations, evolutionary development, and deployment.
- *Roles:* Sorted into business, technical, and management areas.
- *Practices:* Facilitated workshops, prioritizing work using MoSCoW technique, and modeling for enhanced communication.

4. DevOps:

Bridging the traditional chasm between development and operations, DevOps champions holistic software life cycle management.

- *Integration:* Merges development and operations, aiming for faster software delivery and higher quality.
- *Approach:* Teams feel ownership of the entire software life cycle, fostering a sense of responsibility and collaboration.
- *Principle:* "Infrastructure as code" - manage operating environments like how code is managed.
- *Release Strategy:* Frequent, small releases.

5. Crystal:

Crystal's adaptability showcases that not all projects are created equal, and methodologies should reflect this.

- *Variability:* Offers different methodologies depending on team size and software criticality.
- *Crystal Clear:* Resembles Agile and is known for its reliance on trust and communication.
- *Flexibility:* Unlike rigid methodologies, Crystal accommodates individual preferences and habits.

6. Lean Software Development:

Drawing inspiration from lean manufacturing, this approach optimizes efficiency across the software development process.

- *Origin:* Principles from lean manufacturing are applied to software development.
- *Key Principles:*
    1. Eliminate waste.

2. Amplify learning.
3. Deliver fast.
4. Decide late.
5. Empower the team.
6. Build integrity in.
7. Optimize the whole product.

7. Test Driven Development (TDD):

Venturing into the realms of Extreme Programming, we encounter the principle of Test Driven Development, often abbreviated as TDD. At its core, TDD is a proactive and anticipatory approach to software development.

- *Philosophy:* Before you even begin crafting your product, you've already outlined the tests and criteria it must pass. It's the epitome of "beginning with the end in mind."
- *Cross-disciplinary Application:* While primarily rooted in software development, TDD can be harnessed in the realm of physical product creation. By setting up test harnesses or defining validation criteria ahead of time, developers and manufacturers gain a clear blueprint of what the final product should accomplish.
- *Process Overview:* The beauty of TDD lies in its simplicity:
    - **Red:** Initiate by writing the test cases. On execution, these tests are bound to fail because the actual product hasn't been developed yet. This step provides a reference for what is lacking.
    - **Green:** With the red tests in place, begin developing the product. The objective? Ensure all the previously failing tests now pass. This iterative approach ensures that the product meets its intended requirements.
    - **Clean:** Once all tests pass, it's not about resting on laurels. This phase involves refining the product, eradicating redundancies, and ensuring

everything is in compliance with stipulated standards. The outcome is a polished product unit that not only fulfills its requirements but is also efficiently structured.

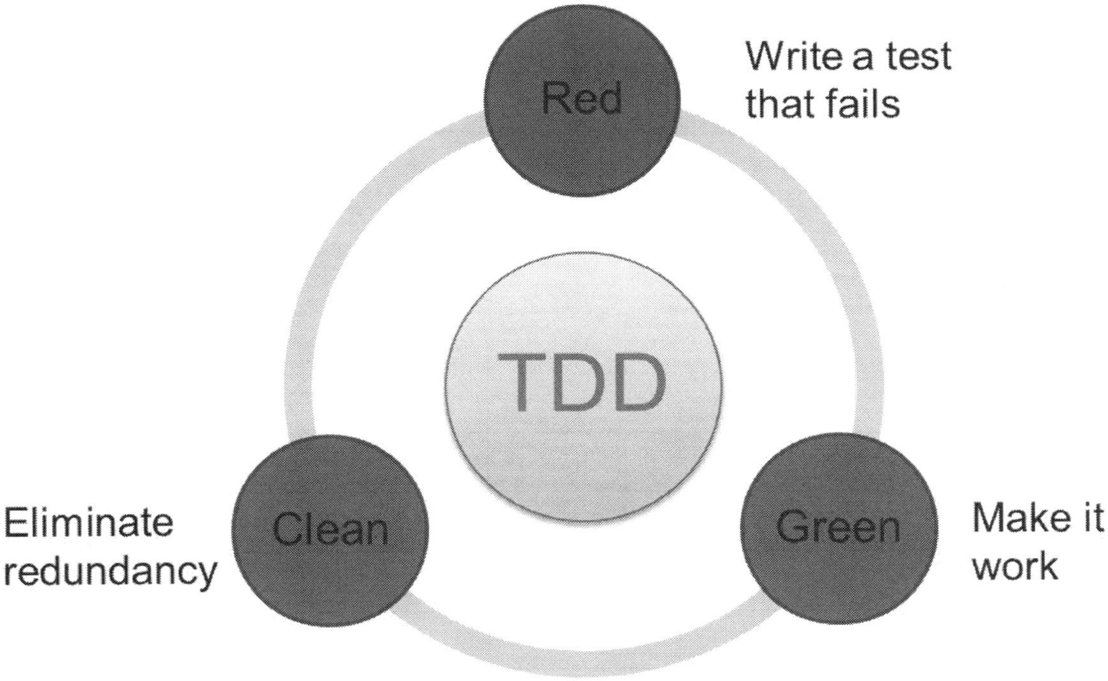

Figure 30: Test Driven Development

The appeal of TDD lies in its systematic approach, which paves the way for efficient and robust product development. By keeping validation criteria in focus right from the outset, it ensures that developers stay aligned with product goals, leading to higher quality and more reliable outcomes.

Closing Thoughts:

The Agile umbrella houses a spectrum of frameworks, each offering unique practices and perspectives. The industry's adaptability ensures it remains relevant as business needs and challenges evolve. It's essential to understand the fundamentals and nuances of each framework to select the right fit for a given project or organizational culture. The

Agile Assessment Guide offers a comprehensive examination of these frameworks, serving as a valuable resource for professionals.

CONTINUOUS INTEGRATION AND CONTINUOUS DEPLOYMENT

Navigating further into the sphere of software development, two concepts demand our attention - Continuous Integration and Continuous Deployment, collectively termed as CI/CD.

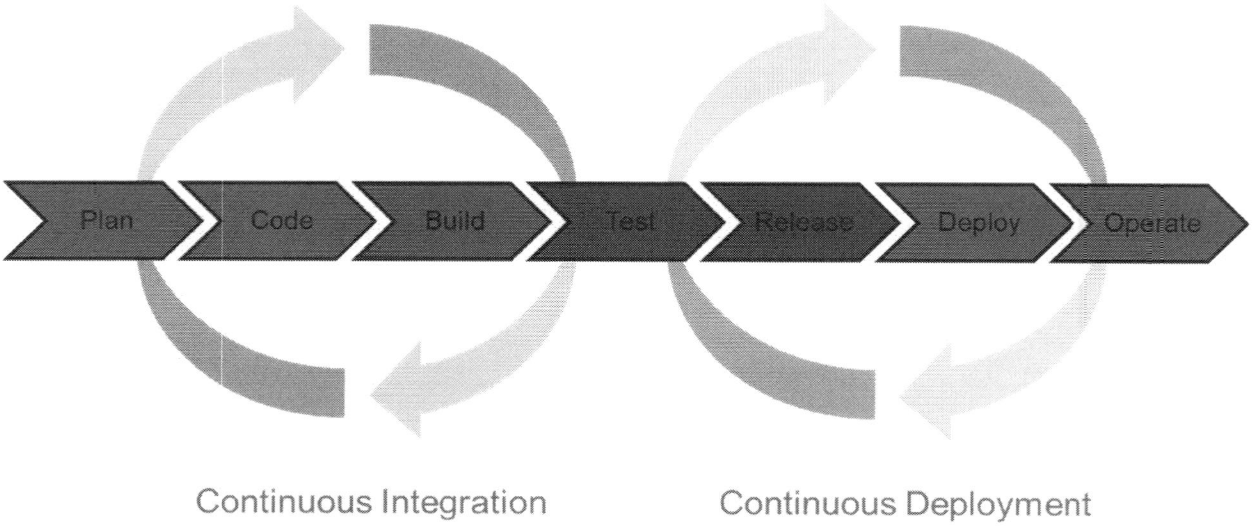

Figure 31: Continuous Integration & Continuous Deployment

Applicability Beyond Software: Although primarily a cornerstone in the software world, the tenets of CI/CD can be transferred to physical product development. It's all about an uninterrupted cycle of integrating changes and swiftly deploying them.

The Breakneck Pace of Integration: To grasp the extremities of CI/CD, one needs to look no further than giants like Amazon. Astonishingly, they integrate changes and push products to production every 6 seconds! While this velocity might seem surreal and isn't necessarily the benchmark for every organization, it exemplifies the potent potential of CI/CD.

The Flow of CI/CD: CI/CD is a harmonious dance of two key steps:

- **Integration:** Incorporate changes into the main product, ensuring they mesh seamlessly and don't disrupt the existing framework.
- **Deployment:** Post integration, swiftly deliver the updated product to its final destination – be it a live environment or into the hands of the customer.

Beyond Just CI/CD: The philosophy of CI/CD often transcends to concepts like DevOps and DevSecOps. These notions encapsulate development, security, and operations into one cohesive pipeline, emphasizing the intertwining of these domains for optimal product delivery.

In Essence: CI/CD is not merely about speed; it's about consistency, quality, and frequent value delivery. By ensuring that every change is immediately integrated, validated, and deployed, CI/CD guarantees a product that's always at its best version when reaching the end-user. And in a rapidly evolving digital landscape, such practices are invaluable for staying ahead of the curve.

QUALITY ASSURANCE

Before we wrap up our discussion, let's explore an area that is often misconceived: Quality Assurance, often abbreviated as QA.

- *Quality - A Mindset, Not Just a Department:* Historically, QA has been seen as the gatekeepers of quality, a separate entity whose sole purpose is to find flaws. However, a modern and more effective perspective sees quality as an inherent attribute, a mindset that every team member should imbibe, rather than a checkpoint at the end of the process.

- *Starting Right at the Genesis:* Quality doesn't magically manifest during the final stages; it's cultivated right from the inception. Whether it's comprehending requirements, drafting designs, or initiating development, ensuring quality at every step minimizes the chances of defects cropping up later.
- *Integrated Quality Control:* In an evolved development ecosystem, the 'QA folks' aren't an isolated unit but an integral part of the entire team. They collaborate from the get-go, aiding in preempting potential pitfalls rather than just spotting them at the end.
- *Striving for Zero:* While no process is absolutely flawless, the aspiration should always be to find zero defects during the final validation. This doesn't negate the importance of post-development validation but stresses on the fact that the more robust the initial stages are, the fewer glitches you'll find later.

In a Nutshell: Quality Assurance isn't just about assurance; it's about instilling a culture of quality. By ensuring that every team member, regardless of their role, is focused on delivering quality at every step, we elevate the entire development process. Remember, quality isn't an act; it's a habit.

WHEN TO TAILOR

Finally, let's review the concept of tailoring. Tailoring is essentially changing the process. Adding steps, removing steps, things like that. Be very, very careful with this. Make sure that you understand what you're doing and why you're doing it before you start tailoring. The concept that we like to use in the Agile community comes from actually martial arts. The concept is called Shu-Ha-Ri. It's a learning process. Three steps to understanding.

Shu is your is your apprentice stage. This is where you don't really understand why we're doing things yet but we're doing them, right? So in this stage, you obey the rules, follow the practice, protect the process, and you're going to protect yourselves, right? And the reason is you don't know yet why we do things so changing them can be a real problem. I equate this to the movie Karate Kid, you know, Wax On Wax Off. He didn't know why he was doing it but when Mr Miyagi started attacking him, he was able to use those practices. Silly reference but you get the idea. Follow the rules when you first start.

Next stage is the Journeyman Stage. This is the Ha stage. This is where you now understand why we do things. You understand the concepts and the core principles. So now you can start detaching from those behaviors or from those rules. You can start experimenting and trying different things. This is a good stage to be. This is where your team is approaching maturity and they understand why we do stuff.

The last stage is Mastery. This is the RI stage. This is where you are just doing it. The practices are habit. They're not something you're thinking about and you know when you're just doing it unconsciously. It's just part of your being, so this is a great stage to be in because the practices don't matter anymore. It's just doing things because they make sense. Because they're the right things to do. So just be careful. Don't tailor too soon or you're likely to violate some of the good stuff that we do and now, you're going to cause harm, okay? Learn the reasons why we do things, tailor over time.

CONCLUSION

As we come to the end of "Agile Principle Raw and Uncut," it's clear that we've covered a lot of ground. For those new to Agile, this book serves as an excellent resource for passing basic Agile exams like the PMI ACP, PMP, CAPM, and others. However, it's important to note that the Scrum Guide is also an essential study material that provides a great level of precision about Scrum in particular. Also check out the Scrum open assessments on www.scrum.org for more practice.

Agile is not a methodology; it's a **mindset** that's backed up by the Agile Manifesto values and principles. As we've seen, there are a variety of Agile practices and frameworks, including Scrum, Kanban, and others. But on the exams from the PMI, the focus is more on an Agile mindset and approach, rather than the rigor of the frameworks and processes.

Agile isn't a methodology, it's a mindset. It's a way of thinking about work that emphasizes collaboration, flexibility, and continuous improvement. It is a way of adapting to the world around you. This mindset is key to success in Agile product development, and it can even be applied to predictive projects.

One anecdote that illustrates the power of the Agile mindset comes from the world of software development. In his book "The Lean Startup," Eric Ries tells the story of how the software company Intuit used Agile principles to create a new version of their QuickBooks software. By adopting an Agile mindset and focusing on customer feedback, they were able to launch a successful product that exceeded their customers' expectations.

In conclusion, Agile is not a set of processes or frameworks. It's a way of thinking about work that emphasizes collaboration, flexibility, and continuous improvement. By keeping an Agile mindset and applying the principles of the Agile Manifesto, you can achieve success in any project, whether it's predictive or Agile. Keep learning and practicing Agile, and you'll be well on your way to becoming an Agile expert. Best wishes on your Agile journey!

Agile Questions Aligned to the Agile Practice Guide

#	Question
1	The first principle of agile places what as the highest priority? A. The Team B. Stakeholders C. Customer Satisfaction D. Time Boxing
2	Agile approaches were created to (select two) A. Explore feasibility in short cycles B. Reduce cost of projects C. Adapt quickly based on evaluation and feedback D. Improve projects with a low level of uncertainty
3	Within the Agile Manifesto, it is stated that we prefer ________________ over processes and tools. A. Individuals and interactions B. Working software C. Customer collaboration D. Responding to change
4	Agile and Kanban can be considered as descendants of: A. Project Management B. Software Development C. Lean Thinking D. Six Sigma
5	When teams deliver ____________, they are better able to understand the true customer requirements faster and more accurately than with a static written specification. A. Working Software B. Small Increments C. Frequently D. Value
6	Which of the following is NOT one of the four types of lifecycles as defined by PMI? A. Waterfall B. Iterative C. Incremental D. Agile
7	True or False: All projects have the characteristics of Requirements, Activities, Delivery, Goal A. True B. False
8	___________ life cycles expect to take advantage of high certainty around firm requirements, a

	stable team, and low risk. A. Predictive B. Incremental C. Iterative D. Agile
9	___________ life cycles improve the product or result through successive prototypes or proofs of concept. Each new prototype yields new stakeholder feedback and team insights. A. Predictive B. Incremental C. Iterative D. Agile
10	When taking an Incremental approach, the characteristics with respect to Delivery can be described as: A. Single Delivery B. Frequent Smaller Deliveries C. Small Deliveries D. Extremely Small Deliveries
11	With respect to the characteristics of Agile Life Cycles, PMI provides two possible ways to achieve incremental delivery so the project aligns with customer needs and can be adapted as necessary. What are those two? A. Scrum B. Iteration-Based Agile C. Kanban D. Flow-Based Agile
12	Agile suitability filter tools A. Help assess under what circumstances an agile approach is appropriate to use B. Help determine the stages of a project C. Establish the best duration for iterations D. Help determine the stages of a Kanban
13	A combination of predictive, iterative, incremental, and/or agile approaches is known as a ________ approach. A. Composite B. Matrix C. Hybrid D. Mixed Mode
14	Using both a predictive and agile approach on the same project is a common scenario. A. True B. False
15	You are managing a construction project. As part of the scope of the project, you are being asked to use a new roofing material. you plan for some small-scale installation trials on the ground first to determine the best installation method and to uncover issues early while there is plenty of time

	to solve them and incrementally improve processes through experimentation and adaptation. This would be described as: A. Agile B. Largely Agile Approach with a Predictive Component C. Predictive D. Predominantly Predictive Approach with Some Agile Components
16	You are working on a project to build a new application. Rather than developing in an incremental way, the decision has been made to simply purchase a third-party product that performs one particular desired capability of the overall application. A. Agile B. Largely Agile Approach with a Predictive Component C. Predictive D. Predominantly Predictive Approach with Some Agile Components
17	You are managing a campus construction project that has multiple buildings to improve and build. You are leveraging an approach that focus resources on completing some buildings earlier than others, thus accelerating the return on investment. This approach would be an example of A. Largely Agile Approach with a Predictive Component B. Predominantly Predictive Approach with Some Agile Components C. Hybrid Life Cycle as Fit-For-Purpose D. Hybrid Life Cycles as Transition Strategy
18	You work for a rather large organization. The CEO would like to implement agile within the organization. Up to this point, your organization has supported a predictive environment. Which of the following approaches would you recommend to begin this transformation? A. Create the roles of Scrum Master and Product Owner and require all project to apply either Kanban or Scrum B. Add more iterative techniques to improve learning and alignment among teams and stakeholders C. Add more incremental techniques to accelerate value and return on investment to sponsors D. Recommend continuing with the predictive approach that the organization has always followed
19	Your team has been working together on a project for some time. You have been leveraging a Kanban board to help visualize and limit the flow of work. In an effort to improve team performance, it has been decided to implement the Scrum framework. This would best be described as: A. Scrum (Cadence Based) B. Kanban (Flow Based) C. Scaling D. Mixing Agile Approaches
20	Which of the following project factors may result in the need for tailoring? A. The flow of work is often interrupted by various delays or impediments B. The quality of the product increments is poor

	C. More than one team is needed to build a product D. All of the above
21	The first step in managing a project using an agile approach requires that the project team _________. A. Adopt an Agile Mindset B. Choose the length of the iteration C. Identify a Scrum Master D. Create a Product Backlog
22	When beginning an agile project, answering the following question will help to develop an implementation strategy: A. How can the project team act in an agile manner? B. What can the team deliver quickly and obtain early feedback to benefit the next delivery cycle? C. How can the team act in a transparent manner? D. All of the above will help
23	Servant Leaders approach work in this order: A. People, Purpose, Process B. Process, People, Purpose C. Purpose, People, Process D. People, Process, Purpose
24	As a Servant Leader, you are responsible for A. Managing relationships to build communication and coordination B. Removing impediments and facilitating the team to streamline its processes C. Assist in fulfilling the team's needs D. All of the above
25	According to PMI, the role of the project manager in an agile project is A. Known as the Scrum Master B. Somewhat of an unknown C. Known as the Product Owner D. Known as the Agile Coach
26	As a servant leader, project managers encourage the distribution of responsibility to A. The Team B. The Product Owner C. The Scrum Master D. The Project Sponsor
27	A core tenet in both the values and the principles of the Agile Manifesto is A. Being a Servant Leader B. The importance of individuals and interactions C. Maintaining a Burndown Chart D. The importance of process and tools
28	An example of how successful agile teams work to collaboratively is:

	A. Pairing B. Swarming C. Mobbing D. All of the above
29	According to the Agile Practice Guide, in agile, three common roles are used. What are the three? A. Cross-functional team members B. Product Owner C. Scrum Master D. Team Facilitator
30	Many successful agile teams are made up of generalizing specialists. Another term to describe this is A. "T-shaped" people B. "I-shaped" people C. Subject Matter Experts D. Developers
31	Distributed teams may have each team member working in a completely different location, either in an office or from home. A. True B. False
32	The key problem with having someone invest only a capacity of 25% or 50% on the team is: A. They are not available when needed B. They will multitask and task switch C. They are not fully committed to the teams goals D. Other team members will not value their opinion
33	Some techniques to consider for managing communication in dispersed teams are A. Fishbowl Windows B. Remote Pairing C. All of the above D. None of the above
34	According to the Agile Practice Guide, the best place to start when forming agile teams is A. Schedule a project kickoff meeting B. Identify the Product Owner C. Build a foundational trust D. Develop a Team Charter
35	According to the Agile Practice Guide, every project needs a A. Project Charter B. Vision Statement C. Product Roadmap D. Team Charter
36	Two questions that should be answered in an agile project charter (select two) A. What is the scope?

	B. Why are we doing this project? C. When should this project be complete? D. Who benefits and how?
37	All of the following are common agile practices, except A. Backlog Refinement B. Daily Standups C. Lessons Learned D. Retrospectives
38	A retrospective allows the team to A. Learn about its process B. Improve its process C. Adapt its process D. All of the above
39	The backlog is the ordered list of all the work, presented in story form, for a team. A. True B. False
40	The product owner is working with the team to prepare some stories for the upcoming iteration and the servant leader is facilitating. What is this practice called? A. Backlog Preparation B. Backlog Refinement C. Iteration Planning D. Daily Standup
41	What are the three questions that should be answered during a Daily Standup? (select three) A. What did I complete since the last standup? B. What did I complete yesterday? C. What am I planning to complete between now and the next standup? D. What are my impediments?
42	As a general guideline, demonstrate whatever the team has as a working product at least once every A. Day B. Week C. 2 weeks D. Month
43	Teams estimate what they can complete each iteration. This is based upon the teams A. Rough Order of Magnitude (ROM) estimate B. Velocity C. Capacity D. Story Points
44	The following technical practices may help the team to deliver at their maximum speed (select all that apply) A. Continuous Integration

	B. Test at all levels C. Test Driven Development (TDD) D. Spikes
45	Consistent iterations help a team establish A. Cadence B. Synchronization C. Unity D. The MVP
46	Agile approaches were born out of the need to A. Solve issues associated with high rates of change B. Uncertainty C. Complexity D. All of the above
47	To effectively measure progress on an agile project, it is best to use A. A traffic light status B. Percent Complete C. Finished work/features D. Earned Value
48	Sponsors usually want to know when the project will be done. Once the team establishes a reliable ________ the team can predict how much longer the project will take. A. Capacity B. Velocity C. Story Point Estimate D. Release Plan
49	Which item influence both the direction and the outcome of any project A. Organization Culture B. Organization Structure C. Organization Policies D. All of the above
50	Which item best describes Organizational Change Management A. It's all about the people B. This is another term for the project change control process C. It covers the skills and techniques for influencing changes that support agility D. All of the above
51	What are the two key factors that further motivate the use of change management practices in an agile context (select two) A. Changes associated with accelerated delivery B. Changes in the organization structure C. Changes associated with agile approaches D. Changes in the business objectives

52	Organizations beginning to use agile approaches should understand the relative compatibility of those methods with their current approaches. Some organizations will have characteristics that more easily support agile principles. All of the following are examples of "friendly" characteristics, except A. A mindset that the organization is leaders in their industry and there is no reason to change B. Executive management's willingness to change C. The organization's willingness to shift the way it views employees D. Focus on short-term budgeting and metrics versus long-term goals
53	Although agile approaches fit well with the lean startup culture, a highly predictive organization can become more agile by A. Encouraging empirical measurements B. Conducting small experiments C. All of the above D. None of the above
54	The most important cultural norm in an organization willing to try any new method or technique is A. To establish a vision B. Enabling a safe work environment C. Creating self-organizing teams D. Creating servant leaders
55	Understanding the organization and the industry requirements that an organization needs to satisfy allows for A. Choosing the right conversations B. Choosing the right tradeoffs C. Choosing the right techniques D. All of the above
56	The Agile Manifesto states "customer collaboration over contract negotiation." This means: A. There will be no contracts on an agile project B. We should enter into Time and Material type contracts only C. We should pursue a shared-risk-reward relationship with our suppliers D. We should enter into Fixed Price type contracts only
57	The willingness and ability to create new competences within an organization when the need arises is a mark of A. Organizational Agility B. Business Agility C. An Agile Mindset D. Business Value
58	Many projects incur dependencies, even when they are not managed within a program. A. True B. False
59	For initiatives that require the collaboration of multiple agile teams in a

	program or portfolio, which approach would be best? A. Scrum B. eXtreme Programming C. Scaled Agile Framework D. Kanban
60	The goal of large-scale agile projects is to A. Coordinate the efforts of different teams to bring value to the business B. Deliver early and often C. Create a Minimal Viable Product D. Coordinate the efforts of different teams to bring value to customer
61	The PMO exists to A. Provide standardized processes and procedures for the organization B. Shepherd business value throughout the organization C. Provide standardized tools and templates for the organization D. Support project managers to enable successful project delivery
62	An agile based PMO A. Enables the goal of delivering the right value, to the right audience, at the right time B. Is based on a customer collaboration mindset C. Functions as if it were a consulting business D. All of the above
63	According to PMI, an agile PMO is "Invitation Oriented". This means that A. Only those individuals or groups who are invited by the PMO have access to the PMO tools and templates B. To drive engagement, the PMO should only invite those that are interested to engage C. Only simple standards will be encouraged D. A recurring meeting will be scheduled to share information with stakeholders who support the PMO
64	An agile PMO should only focus on aspects related to project management A. True B. False
65	The structure of an organization strongly influences its ability to pivot to new information or shifting market needs. Which of the following is NOT a key characteristic of organizational structure? A. Geography B. Size of project deliverable C. Priority of the project D. Allocation of people to projects
66	When implementing a new hybrid or agile approach, it is recommended to undertake the work incrementally. A common tool to help implementing this change is A. Kanban board B. Vision Statement

	C. Product Backlog D. Product Roadmap
67	Adoption and the desire to operate with an agile mindset is limited by the size of the organization or the fact that the organization specializes only in information technology. A. True B. False

Agile Practice Guide Quiz Answers

#	Task	Question
1	AN INTRODUCTION TO AGILE .2	The first principle of agile places what as the highest priority? A. The Team B. Stakeholders **C. Customer Satisfaction** D. Time Boxing
2	2.1 Definable Work vs. High-Uncertainty Work .7	Agile approaches were created to (select two) **A. Explore feasibility in short cycles** B. Reduce cost of projects **C. Adapt quickly based on evaluation and feedback** D. Improve projects with a low level of uncertainty
3	2.2 The Agile Manifesto and Mindset .8	Within the Agile Manifesto, it is stated that we prefer ____________________ over processes and tools. **A. Individuals and interactions** B. Working software C. Customer collaboration D. Responding to change
4	2.3 Lean and the Kanban Method .12	Agile and Kanban can be considered as descendants of: A. Project Management B. Software Development **C. Lean Thinking** D. Six Sigma
5	2.4 Uncertainty, Risk, and Life Cycle Selection .13	When teams deliver _____________, they are better able to understand the true customer requirements faster and more accurately than with a static written specification. A. Working Software **B. Small Increments** C. Frequently D. Value
6	3. LIFE CYCLE SELECTION 17	Which of the following is NOT one of the four types of lifecycles as defined by PMI? **A. Waterfall** B. Iterative C. Incremental D. Agile
7	3.1 Characteristics of Project Life Cycles 18	True or False: All projects have the characteristics of Requirements, Activities, Delivery, Goal **A. True**

		B. False
8	3.1.1 Characteristics of Predictive Life Cycles.20	___________ life cycles expect to take advantage of high certainty around firm requirements, a stable team, and low risk. **A. Predictive** B. Incremental C. Iterative D. Agile
9	3.1.2 Characteristics of Iterative Life Cycles .21	___________ life cycles improve the product or result through successive prototypes or proofs of concept. Each new prototype yields new stakeholder feedback and team insights. A. Predictive B. Incremental **C. Iterative** D. Agile
10	3.1.3 Characteristics of Incremental Life Cycles .22	When taking an Incremental approach, the characteristics with respect to Delivery can be described as: A. Single Delivery **B. Frequent Smaller Deliveries** C. Small Deliveries D. Extremely Small Deliveries
11	3.1.4 Characteristics of Agile Life Cycles .24	With respect to the characteristics of Agile Life Cycles, PMI provides two possible ways to achieve incremental delivery so the project aligns with customer needs and can be adapted as necessary. What are those two? A. Scrum **B. Iteration-Based Agile** C. Kanban **D. Flow-Based Agile**
12	3.1.5 Agile Suitability Filters .25	Agile suitability filter tools **A. Help assess under what circumstances an agile approach is appropriate to use** B. Help determine the stages of a project C. Establish the best duration for iterations D. Help determine the stages of a Kanban
13	3.1.6 Characteristics of Hybrid Life Cycles 26	A combination of predictive, iterative, incremental, and/or agile approaches is known as a ________ approach. A. Composite B. Matrix **C. Hybrid** D. Mixed Mode
14	3.1.7 Combined Agile and Predictive	Using both a predictive and agile approach on the same project is a common scenario.

	Approaches .27	**A. True** B. False
15	3.1.8 Predominantly Predictive Approach with Some Agile Components 28	You are managing a construction project. As part of the scope of the project, you are being asked to use a new roofing material. you plan for some small-scale installation trials on the ground first to determine the best installation method and to uncover issues early while there is plenty of time to solve them and incrementally improve processes through experimentation and adaptation. This would be described as: A. Agile B. Largely Agile Approach with a Predictive Component C. Predictive **D. Predominantly Predictive Approach with Some Agile Components**
16	3.1.9 A Largely Agile Approach with a Predictive Component 28	You are working on a project to build a new application. Rather than developing in an incremental way, the decision has been made to simply purchase a third-party product that performs one particular desired capability of the overall application. A. Agile **B. Largely Agile Approach with a Predictive Component** C. Predictive D. Predominantly Predictive Approach with Some Agile Components
17	3.1.10 Hybrid Life Cycles as Fit-For-Purpose .29	You are managing a campus construction project that has multiple buildings to improve and build. You are leveraging an approach that focus resources on completing some buildings earlier than others, thus accelerating the return on investment. This approach would be an example of A. Largely Agile Approach with a Predictive Component B. Predominantly Predictive Approach with Some Agile Components **C. Hybrid Life Cycle as Fit-For-Purpose** D. Hybrid Life Cycles as Transition Strategy
18	3.1.11 Hybrid Life Cycles as Transition Strategy .30	You work for a rather large organization. The CEO would like to implement agile within the organization. Up to this point, your organization has supported a predictive environment. Which of the following approaches would you recommend to begin this transformation? A. Create the roles of Scrum Master and Product Owner and require all project to apply either Kanban or Scrum **B. Add more iterative techniques to improve learning and alignment among teams and stakeholders** C. Add more incremental techniques to accelerate value and return on investment to sponsors D. Recommend continuing with the predictive approach that the organization has always followed
19	3.2 Mixing Agile Approaches .31	Your team has been working together on a project for some time. You have been leveraging a Kanban board to help visualize and limit the flow of work. In an effort to improve team performance, it has been decided to

		implement the Scrum framework. This would best be described as: A. Scrum (Cadence Based) B. Kanban (Flow Based) C. Scaling **D. Mixing Agile Approaches**
20	3.3 Project Factors That Influence Tailoring .32	Which of the following project factors may result in the need for tailoring? A. The flow of work is often interrupted by various delays or impediments B. The quality of the product increments is poor C. More than one team is needed to build a product **D. All of the above**
21	4. IMPLEMENTING AGILE: CREATING AN AGILE ENVIRONMENT .33	The first step in managing a project using an agile approach requires that the project team _________. **A. Adopt an Agile Mindset** B. Choose the length of the iteration C. Identify a Scrum Master D. Create a Product Backlog
22	4.1 Start with an Agile Mindset .33	When beginning an agile project, answering the following question will help to develop an implementation strategy: A. How can the project team act in an agile manner? B. What can the team deliver quickly and obtain early feedback to benefit the next delivery cycle? C. How can the team act in a transparent manner? **D. All of the above will help**
23	4.2 Servant Leadership Empowers the Team 33	Servant Leaders approach work in this order: A. People, Purpose, Process B. Process, People, Purpose **C. Purpose, People, Process** D. People, Process, Purpose
24	4.2.1 Servant Leader Responsibilities 34	As a Servant Leader, you are responsible for A. Managing relationships to build communication and coordination B. Removing impediments and facilitating the team to streamline its processes C. Assist in fulfilling the team's needs **D. All of the above**
25	4.2.2 Role of the Project Manager in an Agile Environment 37	According to PMI, the role of the project manager in an agile project is A. Known as the Scrum Master **B. Somewhat of an unknown** C. Known as the Product Owner D. Known as the Agile Coach
26	4.2.3 Project Managers Use Servant	As a servant leader, project managers encourage the distribution of responsibility to

	Leadership .38	**A. The Team** B. The Product Owner C. The Scrum Master D. The Project Sponsor
27	4.3 Team Composition 38	A core tenet in both the values and the principles of the Agile Manifesto is A. Being a Servant Leader **B. The importance of individuals and interactions** C. Maintaining a Burndown Chart D. The importance of process and tools
28	4.3.1 Agile Teams 39	An example of how successful agile teams work to collaboratively is: A. Pairing B. Swarming C. Mobbing **D. All of the above**
29	4.3.2 Agile Roles 40	According to the Agile Practice Guide, in agile, three common roles are used. What are the three? **A. Cross-functional team members** **B. Product Owner** C. Scrum Master **D. Team Facilitator**
30	4.3.3 Generalizing Specialists .42	Many successful agile teams are made up of generalizing specialists. Another term to describe this is **A. "T-shaped" people** B. "I-shaped" people C. Subject Matter Experts D. Developers
31	4.3.4 Team Structures .43	Distributed teams may have each team member working in a completely different location, either in an office or from home. A. True **B. False**
32	4.3.5 Dedicated Team Members .44	The key problem with having someone invest only a capacity of 25% or 50% on the team is: A. They are not available when needed **B. They will multitask and task switch** C. They are not fully committed to the teams goals D. Other team members will not value their opinion
33	4.3.6 Team Workspaces 46	Some techniques to consider for managing communication in dispersed teams are A. Fishbowl Windows B. Remote Pairing **C. All of the above** D. None of the above

34	4.3.7 Overcoming Organizational Silos 47	According to the Agile Practice Guide, the best place to start when forming agile teams is A. Schedule a project kickoff meeting B. Identify the Product Owner **C. Build a foundational trust** D. Develop a Team Charter
35	5. IMPLEMENTING AGILE: DELIVERING IN AN AGILE ENVIRONMENT 49	According to the Agile Practice Guide, every project needs a **A. Project Charter** B. Vision Statement C. Product Roadmap D. Team Charter
36	5.1 Charter the Project and the Team 49	Two questions that should be answered in an agile project charter (select two) A. What is the scope? **B. Why are we doing this project?** C. When should this project be complete? **D. Who benefits and how?**
37	5.2 Common Agile Practices 50	All of the following are common agile practices, except A. Backlog Refinement B. Daily Standups **C. Lessons Learned** D. Retrospectives
38	5.2.1 Retrospectives 50	A retrospective allows the team to A. Learn about its process B. Improve its process C. Adapt its process **D. All of the above**
39	5.2.2 Backlog Preparation 52	The backlog is the ordered list of all the work, presented in story form, for a team. **A. True** B. False
40	5.2.3 Backlog Refinement .52	The product owner is working with the team to prepare some stories for the upcoming iteration and the servant leader is facilitating. What is this practice called? A. Backlog Preparation **B. Backlog Refinement** C. Iteration Planning D. Daily Standup
41	5.2.4 Daily Standups .53	What are the three questions that should be answered during a Daily Standup? (select three) **A. What did I complete since the last standup?**

		B. What did I complete yesterday? **C. What am I planning to complete between now and the next standup?** D. **What are my impediments?**
42	5.2.5 Demonstrations/Reviews .55	As a general guideline, demonstrate whatever the team has as a working product at least once every A. Day B. Week **C. 2 weeks** D. Month
43	5.2.6 Planning for Iteration-Based Agile 55	A. Teams estimate what they can complete each iteration. This is based upon the teams B. Rough Order of Magnitude (ROM) estimate C. Velocity **D. Capacity** E. Story Points
44	5.2.7 Execution Practices that Help Teams Deliver Value .56	The following technical practices may help the team to deliver at their maximum speed (select all that apply) **A. Continuous Integration** **B. Test at all levels** **C. Test Driven Development (TDD)** D. **Spikes**
45	5.2.8 How Iterations and Increments Help Deliver Working Product 57	Consistent iterations help a team establish **A. Cadence** B. Synchronization C. Unity D. The MVP
46	5.3 Troubleshooting Agile Project Challenges 57	Agile approaches were born out of the need to A. Solve issues associated with high rates of change B. Uncertainty C. Complexity **D. All of the above**
47	5.4 Measurements in Agile Projects 60	To effectively measure progress on an agile project, it is best to use A. A traffic light status B. Percent Complete **C. Finished work/features** D. Earned Value
48	5.4.1 Agile Teams Measure Results .61	Sponsors usually want to know when the project will be done. Once the team establishes a reliable ________ the team can predict how much longer the project will take. A. Capacity **B. Velocity**

		C. Story Point Estimate D. Release Plan
49	6. ORGANIZATIONAL CONSIDERATIONS FOR PROJECT AGILITY .71	Which item influence both the direction and the outcome of any project A. Organization Culture B. Organization Structure C. Organization Policies **D. All of the above**
50	6.1 Organizational Change Management 71	Which item best describes Organizational Change Management A. It's all about the people B. This is another term for the project change control process **C. It covers the skills and techniques for influencing changes that support agility** D. All of the above
51	6.1.1 Drivers for Change Management .73	What are the two key factors that further motivate the use of change management practices in an agile context (select two) **A. Changes associated with accelerated delivery** B. Changes in the organization structure **C. Changes associated with agile approaches** D. Changes in the business objectives
52	6.1.2 Readiness for Change 73	Organizations beginning to use agile approaches should understand the relative compatibility of those methods with their current approaches. Some organizations will have characteristics that more easily support agile principles. All of the following are examples of "friendly" characteristics, except **A. A mindset that the organization is leaders in their industry and there is no reason to change** B. Executive management's willingness to change C. The organization's willingness to shift the way it views employees D. Focus on short-term budgeting and metrics versus long-term goals
53	6.2 Organizational Culture 75	Although agile approaches fit well with the lean startup culture, a highly predictive organization can become more agile by A. Encouraging empirical measurements B. Conducting small experiments **C. All of the above** D. None of the above
54	6.2.1 Creating an Environment of Safety 75	The most important cultural norm in an organization willing to try any new method or technique is A. To establish a vision **B. Enabling a safe work environment** C. Creating self-organizing teams D. Creating servant leaders

55	6.2.2 Assessing Culture 75	Understanding the organization and the industry requirements that an organization needs to satisfy allows for A. Choosing the right conversations B. Choosing the right tradeoffs C. Choosing the right techniques **D. All of the above**
56	6.3 Procurement and Contracts .77	The Agile Manifesto states "customer collaboration over contract negotiation." This means: A. There will be no contracts on an agile project B. We should enter into Time and Material type contracts only **C. We should pursue a shared-risk-reward relationship with our suppliers** D. We should enter into Fixed Price type contracts only
57	6.4 Business Practices .79	The willingness and ability to create new competences within an organization when the need arises is a mark of **A. Organizational Agility** B. Business Agility C. An Agile Mindset D. Business Value
58	6.5 Multiteam Coordination and Dependencies (Scaling) .80	Many projects incur dependencies, even when they are not managed within a program. **A. True** B. False
59	6.5.1 Frameworks .80	For initiatives that require the collaboration of multiple agile teams in a program or portfolio, which approach would be best? A. Scrum B. eXtreme Programming **C. Scaled Agile Framework** D. Kanban
60	6.5.2 Considerations .80	The goal of large-scale agile projects is to A. Coordinate the efforts of different teams to bring value to the business B. Deliver early and often C. Create a Minimal Viable Product **D. Coordinate the efforts of different teams to bring value to customer**
61	6.6 Agile and the Project Management Office (PMO) 81	The PMO exists to A. Provide standardized processes and procedures for the organization **B. Shepherd business value throughout the organization** C. Provide standardized tools and templates for the organization D. Support project managers to enable successful project delivery
62	6.6.1 An Agile PMO is Value-Driven 81	An agile based PMO A. Enables the goal of delivering the right value, to the right audience, at the right time

		B. Is based on a customer collaboration mindset C. Functions as if it were a consulting business **D. All of the above**
63	6.6.2 An Agile PMO is Invitation-Oriented 81	According to PMI, an agile PMO is "Invitation Oriented". This means that A. Only those individuals or groups who are invited by the PMO have access to the PMO tools and templates **B. To drive engagement, the PMO should only invite those that are interested to engage** C. Only simple standards will be encouraged D. A recurring meeting will be scheduled to share information with stakeholders who support the PMO
64	6.6.3 An Agile PMO is Multidisciplinary .82	An agile PMO should only focus on aspects related to project management A. True **B. False**
65	6.7 Organizational Structure 83	The structure of an organization strongly influences its ability to pivot to new information or shifting market needs. Which of the following is NOT a key characteristic of organizational structure? A. Geography B. Size of project deliverable **C. Priority of the project** D. Allocation of people to projects
66	6.8 Evolving the Organization 84	When implementing a new hybrid or agile approach, it is recommended to undertake the work incrementally. A common tool to help implementing this change is **A. Kanban board** B. Vision Statement C. Product Backlog D. Product Roadmap
67	7. A CALL TO ACTION .87	Adoption and the desire to operate with an agile mindset is limited by the size of the organization or the fact that the organization specializes only in information technology. A. True **B. False**

PRAIZION AGILE BASICS TEAM EXERCISE

WORKING WITH YOUR TEAM, DISCUSS AND ANSWER THE QUESTIONS

1. Which of the following values is considered the highest in the Agile Manifesto?

a. Individuals and interactions

b. Working software

c. Customer collaboration

d. Responding to change

2. The Agile Manifesto values ________ over processes and tools.

a. Interpersonal relationships

b. Contracts and agreements

c. Customer satisfaction

d. Working software

3. Which of the following is NOT one of the Agile Manifesto's four values?

a. Individuals and interactions

b. Working software

c. Customer satisfaction

d. Comprehensive documentation

4. Which of the following values of the Agile Manifesto prioritizes collaboration between the customer and the development team?

a. Individuals and interactions

b. Working software

c. Customer collaboration

d. Responding to change

5. The Agile Manifesto values __________ over following a plan.

a. Flexibility and adaptability

b. Rigid adherence to a schedule

c. Predetermined outcomes

d. Interpersonal relationships

6. Which of the following values of the Agile Manifesto emphasizes the importance of delivering a working product incrementally?

a. Working software

b. Process improvement

c. Detailed documentation

d. Customer collaboration

7. The Agile Manifesto values __________ over contract negotiation.

a. Customer collaboration

b. Formal agreements

c. Individuals and interactions

d. Responding to change

8. Which of the following values of the Agile Manifesto highlights the importance of teamwork and communication?

a. Individuals and interactions

b. Process documentation

c. Customer satisfaction

d. Working software

9. The Agile Manifesto values __________ over following a rigid plan.

a. Adaptability to change

b. Strict adherence to a schedule

c. Predetermined outcomes

d. Interpersonal relationships

10. Which of the following values of the Agile Manifesto prioritizes the delivery of working software over other aspects of the development process?

a. Working software

b. Process improvement

c. Detailed documentation

d. Customer collaboration

11. The Agile Manifesto values __________ over using specific tools.

a. Interpersonal relationships

b. Contract negotiation

c. Customer satisfaction

d. Responding to change

12. Which of the following values of the Agile Manifesto emphasizes the importance of customer involvement in the development process?

a. Individuals and interactions

b. Working software

c. Customer collaboration

d. Responding to change

13. The Agile Manifesto values __________ over comprehensive documentation.

a. Working software

b. Detailed documentation

c. Customer satisfaction

d. Process improvement

14. Which of the following values of the Agile Manifesto prioritizes adapting to change over following a plan?

a. Responding to change

b. Strict adherence to a schedule

c. Predetermined outcomes

d. Interpersonal relationships

15. The Agile Manifesto values __________ over contract negotiation.

a. Customer collaboration

b. Formal agreements

c. Individuals and interactions

d. Responding to change

16. In an Agile team, the team should prioritize which of the following over adhering to a strict plan?

a. Responding to change

b. Strict adherence to a schedule

c. Predetermined outcomes

d. Interpersonal relationships

17. An Agile team is having difficulty balancing the need for documentation with the need for delivering working software. Which of the following should they prioritize?

a. Comprehensive documentation

b. Working software

c. Interpersonal relationships

d. Process improvement

18. A team is struggling with conflicting priorities between the stakeholders and the development team. Which of the following should they prioritize to resolve the conflict?

a. Customer collaboration

b. Contract negotiation

c. Individuals and intentions

d. Responding to change

19. A team is trying to determine the best way to prioritize tasks and allocate resources. Which of the following should they prioritize?

a. Customer engagement

b. Working software

c. Process development

d. Detailed documentation

20. A team is trying to balance the need for efficiency with the need for creativity. Which of the following should they prioritize?

a. Individuals and interactions

b. Processes and tools

c. Detailed documentation

d. Customer management

21. A team is struggling with balancing the need for speed and the need for quality. Which of the following should they prioritize?

a. Responding to change

b. Strict adherence to a schedule

c. Predetermined outcomes

d. Personal goals

22. A team is trying to balance the need for collaboration with the need for accountability. Which of the following should they prioritize?

a. Customer collaboration

b. Contract negotiation

c. Individuals and tools

d. Avoiding change

23. A team is trying to determine the best approach for solving a complex problem. Which of the following should they prioritize?

a. Individuals and interactions

b. Processes and tools

c. Detailed documentation

d. Customer management

24. A team is trying to balance the need for structure with the need for flexibility. Which of the following should they prioritize?

a. Responding to change

b. Strict adherence to a schedule

c. Predetermined outcomes

d. Sprint planning and backlogs

25. A team is trying to determine the best way to approach a new project with limited resources. Which of the following should they prioritize?

a. Working software

b. Comprehensive documentation

c. Individual skills

d. Process improvement

26. A team is struggling with conflicting priorities between the development team and the stakeholders. Which of the following should they prioritize to resolve the conflict?

a. Customer collaboration

b. Contract negotiation

c. Individuals and intentions

d. Generating change requests

27. A team is trying to balance the need for efficiency with the need for creativity. Which of the following should they prioritize?

a. Individuals and interactions

b. Processes and tools

c. Detailed documentation

d. Team management

28. A team is trying to balance the need for speed and the need for quality. Which of the following should they prioritize?

a. Quality

b. Strict adherence to a schedule

c. Predetermined outcomes

d. Personal concerns

29. A team is trying to determine the best approach for delivering customer value. Which of the following should they prioritize?

a. Customer collaboration

b. Contract negotiation

c. Team management

d. Planning no change

30. A team is trying to balance the need for predictability with the need for adaptability. Which of the following should they prioritize?

a. Strict adherence to a schedule

b. Responding to change

c. Predetermined outcomes

d. Interpersonal relationships

31. A team is trying to determine the best way to prioritize work. Which of the following should they prioritize?

a. Project management processes

b. Contract negotiation

c. Processes and plans

d. Working software

32. A team is trying to determine the best way to prioritize stakeholder needs. Which of the following should they prioritize?

a. Customer satisfaction

b. Contract negotiation

c. Individual inspiration

d. Following a plan

33. A team is trying to balance the need for management control with the need for innovation. Which of the following should they prioritize?

a. Innovation

b. Processes and tools

c. Detailed documentation

d. Management control

34. A team is trying to balance the need for predictability with the need for creativity. Which of the following should they prioritize?

a. Responding to change

b. Strict adherence to a schedule

c. Predetermined outcomes

d. Interpersonal relationships

35. Fairouz and Jen are working on a software project for Team ITSAGO. They are struggling with managing conflicting priorities between stakeholders. Which of the following should they prioritize?

a. Stakeholder management

b. Meeting deadlines

c. Technical excellence

d. Customer collaboration

36. Uche and Shouna are leading a sprint review for Team ITSAGO. They are trying to determine the best way to gather feedback from stakeholders. Which of the following should they prioritize?

a. Formal feedback processes

b. Customer collaboration

c. Surveys and questionnaires

d. Contract negotiation

37. Stacy and Ahndrika are working on a feature for Team ITSAGO. They are struggling with how much documentation is necessary. Which of the following should they prioritize?

a. Comprehensive documentation

b. Working software

c. Following a plan

d. Adherence to the timebox

38. Phill and Ahmad are leading a sprint planning meeting for Team ITSAGO. They are trying to balance the need for predictability with the need for adaptability. Which of the following should they prioritize?

a. Strict adherence to a schedule

b. Adaptability

c. Uncertainty

d. Complexity

39. Stacy and Ahndrika are leading a sprint retrospective for Team ITSAGO. They are trying to determine how to improve the team's performance. Which of the following should they prioritize?

a. Continuous improvement

b. Meeting deadlines

c. Technical excellence

d. Stakeholder satisfaction

40. Mark and Korvin are working on a new feature for Team PRAIZION. They are trying to balance the need for speed with the need for quality. Which of the following should they prioritize?

a. Speed

b. Quality

c. Cost

d. Customer satisfaction

41. Lovelace and Ahndrika are leading a sprint retrospective for Team PRAIZION. They are trying to determine how to improve the team's performance. Which of the following should they prioritize?

a. Continuous improvement

b. Meeting deadlines

c. Technical excellence

d. Stakeholder engagement

42. Phill and Ahmad are working on a feature for Team PRAIZION. They are struggling with balancing the need for predictability with the need for creativity. Which of the following should they prioritize?

a. Predictability

b. Creativity

c. Cost

d. Customer satisfaction

43. Frank and Chandra are leading a sprint planning meeting for Team PRAIZION. They are trying to balance the need for speed with the need for quality. Which of the following should they prioritize?

a. Speed

b. Quality

c. Cost

d. Customer satisfaction

44. Courisse and Taji are working on a new feature for Team PRAIZION. They are trying to balance the need for technical excellence with the need for customer satisfaction. Which of the following should they prioritize?

a. Technical excellence

b. Customer satisfaction

c. Meeting deadlines

d. Stakeholder satisfaction

45. Victor and Fairouz are leading a sprint retrospective for Team PRAIZION. They are trying to determine how to improve the team's performance. Which of the following should they prioritize?

a. Continuous improvement

b. Meeting deadlines

c. Technical excellence

d. Stakeholder satisfaction

46. Jen and Uche are working on a feature for Team PRAIZION on an agile endeavor using SCRUM. They are struggling with balancing the need for predictability with the need for adaptability. Which of the following should they prioritize?

a. Predictability

b. Adaptability

c. Cost

d. Customer satisfaction

47. Shouna and Stacy are leading a sprint planning meeting for Team PRAIZION. They are trying to balance the need for collaboration with the need for independence. Which of the following should they prioritize?

a. Collaboration

b. Independence

c. Cost

d. Customer satisfaction

48. Fairouz and Jen are working on a new feature for Team PRAIZION. They are trying to balance the need for technical excellence with the need for stakeholder satisfaction. Which of the following should they prioritize?

a. Technical excellence

b. Stakeholder satisfaction

c. Meeting deadlines

d. Cost

49. Uche and Shouna are leading a sprint retrospective for Team PRAIZION. They are trying to determine how to improve the team's performance. Which of the following events shoulc they prioritize?

a. Predictive plans

b. Meeting deadlines

c. Technical excellence

d. Sprint retrospective

50. Mark and Ahndrika are leading a daily scrum meeting for Team PRAIZION. They are trying to determine how to improve within the team. Which of the following should they prioritize?

a. Communication

b. Cost

c. Technical excellence

d. Meeting deadlines

PRAIZION AGILE BASICS TEAM EXERCISE (ANSWERS)

WORKING WITH YOUR TEAM, DISCUSS AND ANSWER THE QUESTIONS

1. Which of the following values is considered the highest in the Agile Manifesto?

a. Individuals and interactions
b. Working software
c. Customer collaboration
d. Responding to change

Answer: a. Individuals and interactions

Rationale: The Agile Manifesto states that "we value individuals and interactions over processes and tools." This means that people and their relationships are considered more important than following a strict process or using specific tools. Customer collaboration invariably falls under this mindset.

2. The Agile Manifesto values ________ over processes and tools.

a. Interpersonal relationships
b. Contracts and agreements
c. Customer satisfaction
d. Working software

Answer: a. Interpersonal relationships

Rationale: The Agile Manifesto states that "we value individuals and interactions over processes and tools." This means that people and their relationships are considered more important than following a strict process or using specific tools. Interpersonal relationships here = individuals and interactions. Be aware that on PMI exams, you will likely find the language rather indirect and not a straight Agile Manifesto quotation.

3. Which of the following is NOT one of the Agile Manifesto's four values?

a. Individuals and interactions
b. Working software
c. Customer satisfaction
d. Comprehensive documentation

Answer: d. Comprehensive documentation

Rationale: The four values of the Agile Manifesto are individuals and interactions, working software, customer collaboration, and responding to change. Comprehensive documentation is not one of the values stated in the manifesto.

4. Which of the following values of the Agile Manifesto prioritizes collaboration between the customer and the development team?

a. Individuals and interactions
b. Working software
c. Customer collaboration
d. Responding to change

Answer: c. Customer collaboration

Rationale: A bit easy? The Agile Manifesto values "customer collaboration over contract negotiation," which means that collaboration with the customer is considered more important than negotiating a formal agreement.

5. The Agile Manifesto values __________ over following a plan.

a. Flexibility and adaptability
b. Rigid adherence to a schedule
c. Predetermined outcomes
d. Interpersonal relationships
Answer: a. Flexibility and adaptability
Rationale: The Agile Manifesto values "responding to change over following a plan." This means that being flexible and able to adapt to changes is considered more important than following a rigid plan.

6. Which of the following values of the Agile Manifesto emphasizes the importance of delivering a working product incrementally?

a. Working software
b. Process improvement
c. Detailed documentation
d. Customer collaboration
Answer: a. Working software
Rationale: The Agile Manifesto values "working software over comprehensive documentation." This means that delivering a working product incrementally is considered more valuable than creating comprehensive documentation.

7. The Agile Manifesto values __________ over contract negotiation.

a. Customer collaboration
b. Formal agreements
c. Individuals and interactions
d. Responding to change
Answer: a. Customer collaboration
Rationale: The Agile Manifesto values "customer collaboration over contract negotiation," which means that collaboration with the customer is considered more important than negotiating a formal agreement.

8. Which of the following values of the Agile Manifesto highlights the importance of teamwork and communication?

a. Individuals and interactions
b. Process documentation
c. Customer satisfaction
d. Working software
Answer: a. Individuals and interactions

Rationale: The Agile Manifesto values "individuals and interactions over processes and tools." This means that teamwork and communication are considered more valuable than following a strict process or using specific tools.

9. The Agile Manifesto values __________ over following a rigid plan.

a. Adaptability to change
b. Strict adherence to a schedule
c. Predetermined outcomes
d. Interpersonal relationships

Answer: a. Adaptability to change

Rationale: The Agile Manifesto values "responding to change over following a plan." This means that being flexible and able to adapt to changes is considered more important than following a rigid plan.

10. Which of the following values of the Agile Manifesto prioritizes the delivery of working software over other aspects of the development process?

a. Working software
b. Process improvement
c. Detailed documentation
d. Customer collaboration

Answer: a. Working software

Rationale: The Agile Manifesto values "working software over comprehensive documentation." This means that delivering a working product incrementally is considered more important than creating comprehensive documentation.

11. The Agile Manifesto values __________ over using specific tools.

a. Interpersonal relationships
b. Contract negotiation
c. Customer satisfaction
d. Responding to change

Answer: a. Interpersonal relationships

Rationale: The Agile Manifesto values "individuals and interactions over processes and tools." This means that people and their relationships are considered more important than following a strict process or using specific tools.

12. Which of the following values of the Agile Manifesto emphasizes the importance of customer involvement in the development process?

a. Individuals and interactions
b. Working software
c. Customer collaboration
d. Responding to change

Answer: c. Customer collaboration

Rationale: The Agile Manifesto values "customer collaboration over contract negotiation," which means that collaboration with the customer is considered more important than negotiating a formal agreement.

13. The Agile Manifesto values __________ over comprehensive documentation.

a. Working software
b. Detailed documentation
c. Customer satisfaction
d. Process improvement

Answer: a. Working software

Rationale: The Agile Manifesto values "working software over comprehensive documentation." This means that delivering a working product incrementally is considered more important than creating comprehensive documentation.

14. Which of the following values of the Agile Manifesto prioritizes adapting to change over following a plan?

a. Responding to change
b. Strict adherence to a schedule
c. Predetermined outcomes
d. Interpersonal relationships

Answer: a. Responding to change

Rationale: The Agile Manifesto values "responding to change over following a plan." This means that being flexible and able to adapt to changes is considered more important than following a rigid plan.

15. The Agile Manifesto values __________ over contract negotiation.

a. Customer collaboration
b. Formal agreements
c. Individuals and interactions
d. Responding to change

Answer: a. Customer collaboration

Rationale: The Agile Manifesto values "customer collaboration over contract negotiation," which means that collaboration with the customer is considered more important than negotiating a formal agreement.

16. In an Agile team, the team should prioritize which of the following over adhering to a strict plan?

a. Responding to change
b. Strict adherence to a schedule
c. Predetermined outcomes
d. Interpersonal relationships

Answer: a. Responding to change

Rationale: This is based on the Agile Manifesto principle of "Responding to change over following a plan." Agile teams prioritize being flexible and adapting to changes in requirements, rather than rigidly adhering to a predetermined plan.

17. An Agile team is having difficulty balancing the need for documentation with the need for delivering working software. Which of the following should they prioritize?

a. Comprehensive documentation
b. Working software
c. Interpersonal relationships
d. Process improvement

Answer: b. Working software
Rationale: This is based on the Agile Manifesto principle of "Working software over comprehensive documentation." Agile teams prioritize delivering a working product incrementally, rather than creating extensive documentation.

18. A team is struggling with conflicting priorities between the stakeholders and the development team. Which of the following should they prioritize to resolve the conflict?

a. Customer collaboration
b. Contract negotiation
c. Individuals and intentions
d. Responding to change

Answer: a. Customer collaboration
Rationale: This is based on the Agile Manifesto principle of "Customer collaboration over contract negotiation." Agile teams prioritize collaboration with the customer, rather than negotiating a formal agreement. Note the two variables are stakeholders and development team. Development team is not an option in the answer. Also responding to change is not a listed conflict item, so the best answer is option a.

19. A team is trying to determine the best way to prioritize tasks and allocate resources. Which of the following should they prioritize?

a. Customer engagement
b. Working software
c. Process development
d. Detailed documentation

Answer: b. Working software
Rationale: This is based on the Agile Manifesto principle of "Working software over comprehensive documentation." Agile teams prioritize delivering a working product incrementally, rather than creating extensive documentation. While customer engagement is important, the issue here is prioritizing tasks and allocating resources. The most appropriate response to this problem would be to focus resources on working software.

20. A team is trying to balance the need for efficiency with the need for creativity. Which of the following should they prioritize?

a. Individuals and interactions
b. Processes and tools
c. Detailed documentation
d. Customer management

Answer: a. Individuals and interactions

Rationale: This is based on the Agile Manifesto principle of "Individuals and interactions over processes and tools." Agile teams prioritize people and their relationships, rather than following a strict process or using specific tools. Customer management is not a principle of Agile. Customer collaboration is.

21. A team is struggling with balancing the need for speed and the need for quality. Which of the following should they prioritize?

a. Responding to change
b. Strict adherence to a schedule
c. Predetermined outcomes
d. Personal goals

Answer: a. Responding to change

Rationale: This is based on the Agile Manifesto principle of "Responding to change over following a plan." Agile teams prioritize being flexible and adapting to changes in requirements, rather than rigidly adhering to a predetermined plan.

22. A team is trying to balance the need for collaboration with the need for accountability. Which of the following should they prioritize?

a. Customer collaboration
b. Contract negotiation
c. Individuals and tools
d. Avoiding change

Answer: a. Customer collaboration

Rationale: This is based on the Agile Manifesto principle of "Customer collaboration over contract negotiation." Agile teams prioritize collaboration with the customer, rather than negotiating a formal agreement.

23. A team is trying to determine the best approach for solving a complex problem. Which of the following should they prioritize?

a. Individuals and interactions
b. Processes and tools
c. Detailed documentation
d. Customer management

Answer: a. Individuals and interactions

Rationale: This is based on the Agile Manifesto principle of "Individuals and interactions over processes and tools." Agile teams prioritize people and their relationships, rather than following a strict process or using specific tools.

24. A team is trying to balance the need for structure with the need for flexibility. Which of the following should they prioritize?

a. Responding to change
b. Strict adherence to a schedule
c. Predetermined outcomes
d. Sprint planning and backlogs

Answer: a. Responding to change
Rationale: This is based on the Agile Manifesto principle of "Responding to change over following a plan." Agile teams prioritize being flexible and adapting to changes in requirements, rather than rigidly adhering to a predetermined plan.

25. A team is trying to determine the best way to approach a new project with limited resources. Which of the following should they prioritize?

a. Working software
b. Comprehensive documentation
c. Individual skills
d. Process improvement

Answer: a. Working software
Rationale: This is based on the Agile Manifesto principle of "Working software over comprehensive documentation." Agile teams prioritize delivering a working product incrementally, rather than creating extensive documentation. Process improvement is important but the highest priority is to satisfy the customer with working software.

26. A team is struggling with conflicting priorities between the development team and the stakeholders. Which of the following should they prioritize to resolve the conflict?

a. Customer collaboration
b. Contract negotiation
c. Individuals and intentions
d. Generating change requests

Answer: a. Customer collaboration
Rationale: This is based on the Agile Manifesto principle of "Customer collaboration over contract negotiation." Agile teams prioritize collaboration with the customer, rather than negotiating a formal agreement.

27. A team is trying to balance the need for efficiency with the need for creativity. Which of the following should they prioritize?

a. Individuals and interactions
b. Processes and tools
c. Detailed documentation
d. Team management

Answer: a. Individuals and interactions
Rationale: This is based on the Agile Manifesto principle of "Individuals and interactions over processes and tools." Agile teams prioritize people and their relationships, rather than following a strict process or using specific tools.

28. A team is trying to balance the need for speed and the need for quality. Which of the following should they prioritize?

a. Quality
b. Strict adherence to a schedule
c. Predetermined outcomes
d. Personal concerns

Answer: a. Quality
Rationale: This is based on the Agile Manifesto principle of "Technical Excellence and Good design enhances agility." Quality should always be chosen over speed.

29. A team is trying to determine the best approach for delivering customer value. Which of the following should they prioritize?

a. Customer collaboration
b. Contract negotiation
c. Team management
d. Planning no change

Answer: a. Customer collaboration
Rationale: This is based on the Agile Manifesto principle of "Customer collaboration over contract negotiation." Agile teams prioritize collaboration with the customer, rather than negotiating a formal agreement.

30. A team is trying to balance the need for predictability with the need for adaptability. Which of the following should they prioritize?

a. Strict adherence to a schedule
b. Responding to change
c. Predetermined outcomes
d. Interpersonal relationships

Answer: b. Responding to change
Rationale: This is based on the Agile Manifesto principle of "Responding to change over following a plan." Agile teams prioritize being flexible and adapting to changes in requirements, rather than rigidly adhering to a predetermined plan.

31. A team is trying to determine the best way to prioritize work. Which of the following should they prioritize?

a. Project management processes
b. Contract negotiation
c. Processes and plans
d. Working software

Answer: d. Working software

Rationale: This is based on the Agile Manifesto principle of "Working software over comprehensive documentation." Agile teams prioritize delivering a working product incrementally, rather than creating extensive documentation.

32. A team is trying to determine the best way to prioritize stakeholder needs. Which of the following should they prioritize?

a. Customer satisfaction
b. Contract negotiation
c. Individual inspiration
d. Following a plan

Answer: a. Customer satisfaction

Rationale: This is based on the Agile Manifesto principle of "Our highest priority is to satisfy the customer." Agile teams prioritize collaboration with the customer, rather than negotiating a formal agreement.

33. A team is trying to balance the need for management control with the need for innovation. Which of the following should they prioritize?

a. Innovation
b. Processes and tools
c. Detailed documentation
d. Management control

Answer: a. Innovation

Rationale: This is based on the Agile Manifesto principle of " The best architectures, requirements, and designs emerge from self-organizing teams." Agile teams prioritize innovation, rather than following strict management control.

34. A team is trying to balance the need for predictability with the need for creativity. Which of the following should they prioritize?

a. Responding to change
b. Strict adherence to a schedule
c. Predetermined outcomes
d. Interpersonal relationships

Answer: a. Responding to change

Rationale: This is based on the Agile Manifesto principle of "Responding to change over following a plan." Agile teams prioritize being flexible and adapting to changes in requirements, rather than rigidly adhering to a predetermined plan. Even though interpersonal relationships are important, the

question is looking for which variable of two to prioritize. The closest option to this scenario is "responding to change."

35. Fairouz and Jen are working on a software project for Team ITSAGO. They are struggling with managing conflicting priorities between stakeholders. Which of the following should they prioritize?

a. Stakeholder management
b. Meeting deadlines
c. Technical excellence
d. Customer collaboration

Answer: d. Customer collaboration

Rationale: This is based on the Agile Manifesto principle of "Customer collaboration over contract negotiation." Team ITSAGO should prioritize working closely with the customer to understand their needs and deliver a product that meets those needs, rather than focusing on meeting deadlines or achieving technical excellence. How would the customer feel if they were ignored at the expense of the team's personal standards?

36. Uche and Shouna are leading a sprint review for Team ITSAGO. They are trying to determine the best way to gather feedback from stakeholders. Which of the following should they prioritize?

a. Formal feedback processes
b. Customer collaboration
c. Surveys and questionnaires
d. Contract negotiation

Answer: b. Customer collaboration

Rationale: This is based on the Agile Manifesto principle of "Customer collaboration over contract negotiation." Team ITSAGO should prioritize engaging in direct, ongoing communication with stakeholders to gather feedback and improve the product, rather than relying on formal processes or surveys.

37. Stacy and Ahndrika are working on a feature for Team ITSAGO. They are struggling with how much documentation is necessary. Which of the following should they prioritize?

a. Comprehensive documentation
b. Working software
c. Following a plan
d. Adherence to the timebox

Answer: b. Working software

Rationale: This is based on the Agile Manifesto principle of "Working software over comprehensive documentation." Team ITSAGO should prioritize delivering a working product incrementally, rather than creating extensive documentation.

38. Phill and Ahmad are leading a sprint planning meeting for Team ITSAGO. They are trying to balance the need for predictability with the need for adaptability. Which of the following should they prioritize?

a. Strict adherence to a schedule
b. Adaptability
c. Uncertainty
d. Complexity

Answer: b. Adaptability

Rationale: This is based on the Agile Manifesto principle of "Responding to change over following a plan." Team ITSAGO should prioritize being flexible and adapting to changes in requirements, rather than rigidly adhering to a predetermined plan.

39. Stacy and Ahndrika are leading a sprint retrospective for Team ITSAGO. They are trying to determine how to improve the team's performance. Which of the following should they prioritize?

a. Continuous improvement
b. Meeting deadlines
c. Technical excellence
d. Stakeholder satisfaction

Answer: a. Continuous improvement

Rationale: This is based on the Agile Manifesto principles of" At regular intervals, the team reflects on how to become more effective, then tunes and adjusts its behavior accordingly." It is also based on the principle that states "Continuous attention to technical excellence and good design enhances agility." Team ITSAGO should prioritize continuously improving the team's performance and technical skills, rather than just meeting deadlines or satisfying stakeholders.

40. Mark and Korvin are working on a new feature for Team PRAIZION. They are trying to balance the need for speed with the need for quality. Which of the following should they prioritize?

a. Speed
b. Quality
c. Cost
d. Customer satisfaction

Answer: b. Quality

Rationale: In Scrum, the focus should always be on delivering high-quality software. This is because high-quality software is more likely to meet the needs of the customer and be sustainable in the long run. Speed and cost are secondary concerns.

41. Lovelace and Ahndrika are leading a sprint retrospective for Team PRAIZION. They are trying to determine how to improve the team's performance. Which of the following should they prioritize?

a. Continuous improvement
b. Meeting deadlines
c. Technical excellence
d. Stakeholder engagement

Answer: a. Continuous improvement
Rationale: In Scrum, the goal is always to continuously improve the performance of the team. This includes continuously improving the team's technical skills, as well as its ability to meet deadlines and satisfy stakeholders. For questions like this, be sure to stay focused on the problem at hand. In this question, the problem at hand is team improvement. The other options are not as relevant or are secondary and can be rolled up under continuous improvement.

42. Phill and Ahmad are working on a feature for Team PRAIZION. They are struggling with balancing the need for predictability with the need for creativity. Which of the following should they prioritize?

a. Predictability
b. Creativity
c. Cost
d. Customer satisfaction
Answer: b. Creativity
Rationale: In Scrum, creativity is important in order to find innovative solutions to problems and come up with new ideas that can improve the product. Predictability is important, but it should not be prioritized at the expense of creativity. The question does not mention customer satisfaction as being a factor needing balance, so in this case, it is not the best option since choosing "customer satisfaction" leaves the predictability vs. creativity problem unsolved.

43. Frank and Chandra are leading a sprint planning meeting for Team PRAIZION. They are trying to balance the need for speed with the need for quality. Which of the following should they prioritize?

a. Speed
b. Quality
c. Cost
d. Customer satisfaction
Answer: b. Quality
Rationale: In Scrum, the focus should always be on delivering high-quality software. This is because high-quality software is more likely to meet the needs of the customer and be sustainable in the long run. Speed and cost are secondary concerns.

44. Courisse and Taji are working on a new feature for Team PRAIZION. They are trying to balance the need for technical excellence with the need for customer satisfaction. Which of the following should they prioritize?

a. Technical excellence
b. Customer satisfaction
c. Meeting deadlines
d. Stakeholder satisfaction
Answer: b. Customer satisfaction
Rationale: In Scrum, customer satisfaction is a key priority. The team should strive to deliver a product that meets the needs of the customer, rather than just focusing on technical excellence or meeting deadlines. This is our highest priority in Agile Principle #1

45. Victor and Fairouz are leading a sprint retrospective for Team PRAIZION. They are trying to determine how to improve the team's performance. Which of the following should they prioritize?

a. Continuous improvement
b. Meeting deadlines
c. Technical excellence
d. Stakeholder satisfaction

Answer: a. Continuous improvement

Rationale: The problem at hand here is team improvement. In Scrum, the goal is always to continuously improve the performance of the team. This includes continuously improving the team's technical skills, as well as its ability to meet deadlines and satisfy stakeholders.

46. Jen and Uche are working on a feature for Team PRAIZION on an agile endeavor using SCRUM. They are struggling with balancing the need for predictability with the need for adaptability. Which of the following should they prioritize?

a. Predictability
b. Adaptability
c. Cost
d. Customer satisfaction

Answer: b. Adaptability

Rationale: Again, this question presents the distractor of "customer satisfaction" as an irrelevant red herring. The two variables in question should be focused on instead. In Scrum, adaptability is important because the requirements of the product may change over time. The team should be able to adapt to these changes, rather than being locked into a rigid plan. Predictability is important, but it should not be prioritized at the expense of adaptability.

47. Shouna and Stacy are leading a sprint planning meeting for Team PRAIZION. They are trying to balance the need for collaboration with the need for independence. Which of the following should they prioritize?

a. Collaboration
b. Independence
c. Cost
d. Customer satisfaction

Answer: a. Collaboration

Rationale: Again, this question presents the distractor of "customer satisfaction" as an irrelevant red herring. The two variables in question (collaboration vs. independence) should be focused on instead. In Scrum, collaboration is important because it helps the team to work together effectively and achieve common goals. Independence is important, but it should not be prioritized at the expense of collaboration.

48. Fairouz and Jen are working on a new feature for Team PRAIZION. They are trying to balance the need for technical excellence with the need for stakeholder satisfaction. Which of the following should they prioritize?

a. Technical excellence
b. Stakeholder satisfaction
c. Meeting deadlines
d. Cost

Answer: b. Stakeholder satisfaction
Rationale: This question should be ingrained by now as it has been repeated for emphasis. In Scrum, stakeholder satisfaction is a key priority. The team should strive to deliver a product that meets the needs of stakeholders, rather than just focusing on technical excellence or meeting deadlines.

49. Uche and Shouna are leading a sprint retrospective for Team PRAIZION. They are trying to determine how to improve the team's performance. Which of the following events should they prioritize?

a. Predictive plans
b. Meeting deadlines
c. Technical excellence
d. Sprint retrospective

Answer: d. Sprint retrospective
Rationale: In Scrum, the goal is always to continuously improve the performance of the team. This includes continuously improving the team's technical skills, as well as its ability to meet deadlines and satisfy stakeholders. Tuning and adjusting in scrum is continuous but the specific event to prioritize for this is the Sprint Retrospective.

50. Mark and Ahndrika are leading a daily scrum meeting for Team PRAIZION. They are trying to determine how to improve within the team. Which of the following should they prioritize?

a. Communication
b. Cost
c. Technical excellence
d. Meeting deadlines

Answer: a. Communication
Rationale: In Scrum, effective communication is critical to the success of the team. The team should prioritize communication in order to ensure that everyone is on the same page and working effectively together.

PRAIZION SCRUM TEAM EXERCISE 1

ID	**Questions** Answer the following questions as a team and collaborate to find the answers.	**Answers**
1	How many roles are there in Scrum?	
2	Scrum is a ________________ framework that helps people, teams and organizations generate value.	
3	In Scrum, a __________ orders the work for a complex problem into a Product Backlog	
4	The ____________ turns a selection of the work into an Increment of value during a Sprint.	
5	The ____________ and its __________inspect the results and adjust for the next Sprint.	
6	Scrum has a configuration made up of three numbers which is ___ _____ _____	
TRUE OF FALSE?		
7	The ideal size of an agile team is 6 to 15 people	
8	The product backlog is refined only at the end of the Retrospective	
9	The Product Owner is a member of the Scrum Team	
10	The project stakeholders accept or reject the PSI in the Sprint Retrospective meeting	
11	The purpose of the Sprint Retrospective is to plan ways to increase quality and effectiveness	
12	The Scrum Master can delegate tasks to team members	
13	The Scrum Master is also known as the Assistant Product Owner	
14	The Scrum Team has the power to cancel a troubled sprint	
15	The single most important meeting in Scrum is the Retrospective	
20	The Sprint is a container for all other events	

PRAIZION SCRUM TEAM EXERCISE 2

ID	**Questions** Answer the following questions as a team and collaborate to find the answers.	**Answers**
1	The definition of ready can be understood from the ______ acronym.	
2	__________ criteria pertains to user story acceptance while definition of _________ pertains to acceptance of the Increment.	
3	PSI stands for _________________	
4	In Kanban, one of the tenets espoused is to limit your __________	
5	At the end of each iteration, the team adds up effort estimates associated with user stories that were completed during that iteration. This total is called ___________.	
6	"Start where you are". "Respect the existing roles." Are tenets of _________	
	TRUE OF FALSE?	
7	In Kanban, 2 types of time often discussed are ___________ time and ____________ time.	
8	Time from customer order to customer delivery is _________ time.	
9	Time from picking up an order to customer delivery is _________ time.	
10	Individuals and Interactions OVER________________	
11	Your team can deliver 200 story points in a sprint. This is known as ____________	
12	Tasks are estimated in ____________________	
13	User Stories are estimated in ______________	
14	The 5 stages of team development are ______	
15	1, 2, 3, 5, 8, 13 are numbers used in which estimating approach? _____________	
16	XS, S, M, L, XL, XXL are categories used in which estimating? ______________	
17	Who can add to the backlog?	
18	________________ provide visibility into work remaining	
19	____________ provide visibility into work completed	
20	A project has a low frequency of delivery and a high degree of change. Which delivery method should you use? ___________________	

PRAIZION SCRUM TEAM EXERCISE 3

ID	**Questions** Answer the following questions as a team and collaborate to find the answers.	**Answers**
1	____________ is responsible for managing work progress during a Sprint in Scrum.	
2	In industry, the average length of a Sprint is ___________________	
3	How long could a sprint be? 4 Weeks or 4 Months?	
4	How many people should there be on a Scrum team?	
5	Ordering the Product Backlog is the responsibility of ___________________	
6	Giving status in a daily stand-up meeting and turning it into a stand-up meeting is an example of an ____________	
7	Scrum is founded on emp_________________.	
8	The team has a broad set of skills with deep specialisms. This describes ___________ skills	
9	Customer collaboration OVER ______	
10	How many HYBRID models were discussed in the Agile Practice Guide?	
11	The team has a broad set of skills with varying degrees of specialisms. This describes ___________ comb.	
12	What is the best thing to do with a well-functioning Agile team?	
13	You become the sole owner of a user story at what point in Scrum _________	
14	At the end of a sprint, a team MUST ship an increment to the customer or release to production. True or False.	
15	You are working on a complicated project with a low frequency of delivery and a low degree of change. Which approach should you use? Predictive or change-driven?	
16	Discuss the Scrum Skeleton	

PRAIZION SCRUM TEAM EXERCISE 1 - ANSWERS

ID	**Questions** Answer the following questions as a team and collaborate to find the answers.	**Answers**
1	How many roles are there in Scrum?	3
2	Scrum is a ________________ framework that helps people, teams and organizations generate value.	Lightweight
3	In Scrum, a __________ orders the work for a complex problem into a Product Backlog	Product Owner
4	The ____________ turns a selection of the work into an Increment of value during a Sprint.	Scrum Team
5	The ____________ and its __________inspect the results and adjust for the next Sprint.	Scrum Team, stakeholders
6	Scrum has a configuration made up of three numbers which is ___ ____ ____	3 ...5...3
	TRUE OF FALSE?	
7	The ideal size of an agile team is 6 to 15 people	FALSE
8	The product backlog is refined only at the end of the Retrospective	FALSE
9	The Product Owner is a member of the Scrum Team	TRUE
10	The project stakeholders accept or reject the PSI in the Sprint Retrospective meeting	FALSE
11	The purpose of the Sprint Retrospective is to plan ways to increase quality and effectiveness	TRUE
12	The Scrum Master can delegate tasks to team members	FALSE
13	The Scrum Master is also known as the Assistant Product Owner	FALSE
14	The Scrum Team has the power to cancel a troubled sprint	FALSE
15	The single most important meeting in Scrum is the Retrospective	TRUE
20	The Sprint is a container for all other events	TRUE

PRAIZION SCRUM TEAM EXERCISE 2 - ANSWERS

ID	**Questions** Answer the following questions as a team and collaborate to find the answers.	**Answers**
1	The definition of ready can be understood from the ______ acronym.	INVEST
2	__________ criteria pertains to user story acceptance while definition of _________ pertains to acceptance of the Increment.	Acceptance, Done
3	PSI stands for _________________	Potentially Shippable Increment
4	In Kanban, one of the tenets espoused is to limit your __	WIP
5	At the end of each iteration, the team adds up effort estimates associated with user stories that were completed during that iteration. This total is called ____________.	Velocity
6	"Start where you are". "Respect the existing roles." Are tenets of _________	Kanban
	TRUE OF FALSE?	
7	In Kanban, 2 types of time often discussed are ___________ time and ____________ time.	Lead Time and Cycle Time
8	Time from customer order to customer delivery is _________ time.	Lead Time
9	Time from picking up an order to customer delivery is _________ time.	Cycle Time
10	Individuals and Interactions OVER________________	Processes and tools
11	Your team can deliver 200 story points in a sprint. This is known as ____________	Velocity
12	Tasks are estimated in ____________________	Hours
13	User Stories are estimated in ______________	Story Points
14	The 5 stages of team development are ______	Forming - Storming - Performing - Adjourning
15	1, 2, 3, 5, 8, 13 are numbers used in which estimating approach? _____________	Planning Poker
16	XS, S, M, L, XL, XXL are categories used in which estimating? _______________	Affinity Estimating
17	Who can add to the backlog?	Anyone
18	________________ provide visibility into work remaining	Burndown charts
19	____________ provide visibility into work completed	Burnup charts
20	A project has a low frequency of delivery and a high degree of change. Which delivery method should you use? ___________________	Iterative

PRAIZION SCRUM TEAM EXERCISE 3 - ANSWERS

ID	**Questions** Answer the following questions as a team and collaborate to find the answers.	**Answers**
1	____________ is responsible for managing work progress during a Sprint in Scrum.	Development Team
2	In industry, the average length of a Sprint is __________________	2 Weeks
3	How long could a sprint be? 4 Weeks or 4 Months?	4 weeks
4	How many people should there be on a Scrum team?	10
5	Ordering the Product Backlog is the responsibility of __________________	Product Owner
6	Giving status in a daily stand-up meeting and turning it into a stand-up meeting is an example of an ____________	Anti-pattern
7	Scrum is founded on emp_________________.	empiricism
8	The team has a broad set of skills with deep specialisms. This describes ___________ skills	T-shaped
9	Customer collaboration OVER ______	Contract negotiation
10	How many HYBRID models were discussed in the Agile Practice Guide?	4
11	The team has a broad set of skills with varying degrees of specialisms. This describes ___________ comb.	Broken
12	What is the best thing to do with a well-functioning Agile team?	Keep them together
13	You become the sole owner of a user story at what point in Scrum _________	You are never "sole owner"
14	At the end of a sprint, a team MUST ship an increment to the customer or release to production. True or False.	FALSE
15	You are working on a complicated project with a low frequency of delivery and a low degree of change. Which approach should you use? Predictive or change-driven?	Predictive
16	Discuss the Scrum Skeleton	Explain 3-5-3

AGILE CERTIFIED PRACTITIONER HALF-MARATHON

1. Your senior executives are exhibiting behaviors that are in direct contrast to the agile manifesto. You have historically used a traditional approach and only recently started employing agile frameworks. What should you do as a servant leader/scrum master in the firm?

   a) Ask the product owner to coach the senior executives
   b) Model agile principles and discuss with stakeholders
   c) Send the company for mandatory training with comprehensive documentation
   d) Develop a plan for a thorough rollout with an external training firm

2. Your senior executives are exhibiting behaviors that are in direct contrast to the agile manifesto. You have historically used a traditional approach and only recently started employing agile frameworks. What should you do as a servant leader/scrum master in the firm?

   a) Train the team in Kanban and Scrum
   b) Send the company for mandatory training with comprehensive agile documentation
   c) Explore the possibility of using a predictive method with more structure
   d) Build common knowledge around agile practices and terminology to work effectively

3. Your senior executives are exhibiting behaviors that are in direct contrast to the agile manifesto. You have historically used a traditional approach and only recently started employing agile frameworks. As an agile leader, which of the following is your responsibility?

   a) Support change at the system or organization level
   b) Educate the organization
   c) Influence processes, behaviors, and people
   d) All of the options

4. Your agile team is struggling on the project. Transparency is a huge issue and the team has lost stakeholder trust. Estimates seem to be secrets and real progress cannot be seen by the stakeholders. What is the best course of action?

   a) Use the concept of simplifying acceptance
   b) Set aside between 6 – 8 hours for demonstrations
   c) Use information radiators
   d) Use retrospectives with burndown charts

The other options do not remedy the transparency issue because they are all reactive (after work has occurred in acceptance, demos and retrospectives - where stakeholders would not likely be in attendance anyway.)

5. Having just converted to agile approaches, your team has made a number of mistakes using Scrum and is struggling with a sporadic demand pattern. What is the best thing to do?

   a) Iterative based agile
   b) Strict adherence to established cadence
   c) Flow-based agile
   d) Quality audits to enforce short sprints and no mistakes

6. Having just converted to agile approaches, your team is building a new product. The team may need feedback on a prototype to glean customer feedback to understand what they need to provide for subsequent delivery of the final finished feature . What is the best thing to do?

a) Definition of done
b) Minimum Marketable Increment
c) Minimum Viable Product
d) Marketable Viable Feature

7. Having just converted to agile approaches, your team is involved in the progressive elaboration of project requirements. They are involved in the ongoing activity in which the team collaboratively reviews, updates, and writes requests/requirements to satisfy the need of the customer request. Which action does this best describe?

a) Code authoring and analysis
b) Requirements elicitation
c) Backlog refinement
d) Product road-mapping

8. Having just converted to agile approaches, your team is involved in the progressive elaboration of project requirements. As you proceed, certain individuals appear to naturally have more influence than others. What should you do as the servant leader?

a) Restrict influence to a select few
b) Discourage self-organization promote team empowerment
c) Encourage centralized leadership
d) Encourage emergent leadership

9. Having just converted to agile approaches, your team is involved in the progressive elaboration of project requirements. What should you do as the servant leader?

a) Develop detailed plans and roadmaps for the team
b) Define scope and schedule for the project team
c) Promote collaboration and conversation between teams
d) Encourage rewards for exceptional individuals

10. In agile, when ordering the backlog, it could be done primarily by:

a) Delighters and satisfiers
b) Nice to haves and won't haves
c) Should haves and could haves
d) Value and benefit

11. You are using an archetype user representing a set of similar end users described with their goals, motivations, and representative personal characteristics. What does this best describe?

a) Personas
b) Pairing
c) Fishbowls
d) Story-writing

12. You are a servant leader on an agile project. Towards the end of the second iteration, you discover that a new stakeholder is in attendance at the Sprint review. This is the first time you are encountering this stakeholder. What course of action would most likely have prevented this surprise?

a) Identifying all stakeholders before the first iteration
b) Identifying all the stakeholders before the first Sprint review
c) Identifying all the stakeholders before product visioning
d) Identifying and engaging stakeholders through periodic reviews

13. You work for a Fortune 500 firm in London. Over the past 15 years, you have been a program manager using predominantly predictive project management approaches. Recently, the CEO announced a re-organization and a new agile approach to planning, and you are now an Agile Certified Practitioner and Scrum Master. What is the best way to identify and engage stakeholders?

a) Direct the stakeholders to view low tech, high-touch visual displays
b) Promoting knowledge sharing early and throughout the project
c) Maximizing the executing phase and assigning stakeholders testing tasks to run
d) Send the stakeholders burn-down charts, and status reports, at the end of each release

14. An agile transformation has just taken place in your firm after 4 weeks of training senior management (VPs, Directors and Managers). Recently, the CEO also announced an entire re-organization and this new agile approach. You begin using agile frameworks on your product delivery.
In your third iteration, you realize that stakeholders are not engaged. What should you do first?

a) Report the matter to senior management and get support from higher levels of authority.
b) Apply directional leadership while receiving support from the CEO to influence stakeholders.
c) Apply servant leadership to work with stakeholders and possibly product owner.
d) Invite the stakeholders to as many events as possible to get them engaged an active.

15. An Agile team discovered during the first couple of iterations that there was a gap in the expectations of the stakeholders and the deliverables produced by the team.
Which of the following would be most effective in reducing the gap?

a) Delegate product owner responsibilities to more than one stakeholder to engage them
b) Have the team carry out a stakeholder gap-analysis and generate a comprehensive report
c) Continually assess changes in the organization to ensure stakeholder engagement
d) Assign all story-writing to all new stakeholders to get them as engaged as possible

16. An Agile team needs services of a technical expert to run certain tasks. Some of the team has concerns about this. They feel this arrangement might impede their progress and raised this concern to the Team Facilitator. What is the best course of action for the Team Facilitator?

a) Escalate the matter to Senior Management and HR, then flag it as an impediment
b) Use Highsmith decision spectrum or voting for collaborative team decision making
c) Meet with the technical expert alone in order to establish a working arrangement
d) Allow the product owner and stakeholders to decide upon the outcome for this

17. As a facilitator of an agile team, you are seeking to establish a shared vision of the various project increments with the team and senior management. How should you go about this task?

a) Develop a team charter and sprint backlog to share the vision
b) Develop a team contract and release plan to expand on the vision and share accordingly
c) Develop an agile project charter to share the vision with stakeholders
d) Develop a wire-frame, release plan and sprint goal to share the ideas

18. While working together with your agile team, team members share concerns about a lack of trust among stakeholders. There seems to be an underlying belief that the team will fail. What should you do as a servant leader?

a) Interface with the stakeholders daily to review definition of ready, acceptance criteria and spikes to create a sense of trust.
b) Create a culture of independence from the team by communicating with stakeholders mainly at the sprint review ceremony, backlog preparation and refinement.
c) Deliver as much value as possible every iteration and encourage the team to work overtime to impress the stakeholders early on in the first few sprints.
d) Create an environment with a shared understanding of deliverables, success criteria and acceptable trade-offs.

19. Based on the Agile PMO's advice, your agile team of 45 members, has begun to scale practices being used in the firm. As they do this, they receive a lot of stakeholder pushback. Leadership of the technology group has accused the team of wasting valuable development time in too many scrum-of-scrums. What is the best thing for you to do as the servant leader?

a) Share with them that a scrum of scrums is not a disruptive meeting. In fact, it helps put everyone on the same page, prioritize value, and avoid all hidden risks.
b) Rely on the structure and mediation of the project management office. Report the matter to a senior PMO officer and have them mediate the matter.
c) Mentor stakeholders and be transparent regarding the work process and status by communicating elements of concern, impediments, quality and risks.
d) Provide transparency regarding work status by restricting the flow of unfavorable metrics such as capacity, earned value and velocity that may confuse stakeholders.

20. Your production team has moved from a predictive method to an agile approach. You are using the concepts of work in progress, cost of delay, and cycle time to provide a high-quality product. As the project proceeds, stakeholders ask for status. What should you do?

a) Provide forecasts at a granular level of detail that eliminates uncertainty in the planning process to enable stakeholders plan the entire project.
b) Provide data to balance the need for uncertainty and the benefits of predictability in order to allow stakeholders to plan each epic or task.
c) Give the stakeholders forecasts, at a level of detail that balances benefits of certainty and adaptability so they can effectively plan.
d) Provide forecasts at a high-level of detail that negates certainty. Try as much as possible to stick to original plans without deviation to minimize work.

21. A few weeks ago, you were working on the social contract foster team coherence and strengthen team members' commitment to shared outcomes. A few weeks later, your team is in a heightened conflict stage where you need to refer to this contract. Which stage of team development are they likely in?

a) Performing
b) Storming
c) Norming
d) Avoiding

22. Your team has found themselves in hyper-norming. Everyone agrees about everything even when it is wrong. Which of the following should the servant leader uphold?

a) Constructive disagreement
b) Good guy bad guy
c) Hyper-conflict
d) Theory x Management

23. As a servant leader you are helping to create a team of generalizing specialists that has interpersonal and technical skills needed to achieve the project objectives. What is the main purpose and objective for these actions?

a) To equip the team with as many skills as possible
b) To create business value with minimal delay
c) To create a team with I-shaped profiles
d) To create a team with paint-drip, broken comb or T-shaped profiles

24. Your agile team is performing at its peak. All team members are fully engaged and involved. Team members need little direction and support. Which leadership approach would best suit the team at this stage?

a) Delegating
b) Directing
c) Telling
d) Supporting

25. Your agile team is very fragile and performing poorly. All team members are disengaged and worried. Team members need a lot of direction and support. Which leadership approach would best suit the team at this stage?

a) Delegating
b) Coaching
c) Telling
d) Supporting

26. As the servant leader of an agile team, you observe that a team member has the need for very little direction but a lot of support. What is the best approach for leading in this circumstance?

a) Participating
b) Directing
c) Coaching
d) Delegating

27. Your project team complains to you about certain stakeholders who have breached the team space, making requests for work not included in the sprint. As a self-organizing team, what is your best course of action as a servant leader?

a) Ask the team to solve the problem themselves
b) Shield the team from interruptions
c) Decide to remove the impediments causing their distress
d) Report the matter to Senior Management and hold a meeting

28. Your project team complains to you about certain project and release aspects. As a self-organizing team, what is your best course of action as a servant leader?

a) Ask the team to solve the problem themselves
b) Work with the product owner and stakeholders to find resolve
c) Decide to remove the impediments causing their distress
d) Report the matter to Senior Management and hold a meeting

29. You are a servant leader on an agile team deep in the middle of development. Which of the following would your role most likely be?

a) Working on plans and reports
b) Drawing out the next release and iteration plan
c) Solving personal conflicts issues for the team
d) Carrying food and water

30. Your agile team has put in a lot of work towards a product release. Which of the following best describes an immediate reward system that you would employ for the team at this time?

a) Monetary rewards for employees who have worked the hardest
b) Treating the team to dinner at a five-star resort
c) Allowing each team member to choose what they each prefer on a first come first served basis from a gift bag
d) Awarding the team more challenging work on a future high visibility project

31. Your team was in the same location previously, but now they are virtually co-located. Which of the following would you expect as a result of this colocation?

a) Video cameras and conferencing are used less
b) Conflict resolution between team members is easier
c) Instant messaging would most likely not be needed
d) Greater challenges in conflict resolution

32. While working together with your agile team, team members share concerns about stakeholder's perception of commitment and goal alignment. What should you do as a servant leader?

a) Ensure increasingly detailed levels of requirements in order to ensure common understanding
b) Create a culture of independence by communicating with stakeholders mainly at the sprint review
c) Deliver as much value as possible every iteration
d) Share the project vision and ensure the team understands how their objectives fit the overall goals

33. Which of the following is true about Agile teams?

a) When people are unavailable, the product owner understands that the team has reduced capacity.
b) When people are unavailable, the product owner understands that the team has reduced lead time.
c) When people are unavailable, the product owner understands that the team has reduced cycle time.
d) When people are unavailable, the product owner understands that the team has increased redundancy.

34. As a servant leader, it is important that you employ emotional intelligence. Which of the following best describes emotional intelligence?
   a) Applying the Meyers-Briggs model to every situation before decision making
   b) Being socially aware and kind to people
   c) Controlling your emotions and controlling the emotions of others whichever way possible
   d) Controlling one's emotions and influencing the emotions of other individuals and groups

35. As a team using an agile framework, which of the following approaches to planning would you expect?
   a) Detailed planning at the start of each sprint
   b) Daily planning during the daily scrum
   c) Plan at multiple levels (strategic, release, iteration, daily)
   d) Planning during each sprint and release to maximize value

36. You work for a Fortune 500 firm in London. Recently, the CEO announced a re-organization and a new agile approach to planning, and you are now an Agile Certified Practitioner. What is the best way to make planning activities visible and transparent?
   a) Direct the stakeholders to use hi-tech, high-touch visual displays
   b) Encourage participation of key stakeholders and publishing planning results
   c) Maximizing the planning phase by assigning stakeholders to easy planning tasks
   d) Encourage stakeholders to participate in creating burn-down charts each release

37. Your team is practicing scrum as an agile framework. Which scrum event in your sprint, will give your team the greatest opportunity to inspect and adapt the project plan to reflect changes in requirements, schedule, budget, and shifting priorities?
   a) Daily Scrum
   b) Sprint Review
   c) Backlog Refinement
   d) Retrospective

38. As you consider range estimates on your agile project, which of the following ideas should you take into account?
   a) Ignore maintenance and operations demands this is outside the project
   b) Be sure to not adjust capacity. Do not incorporate other factors or work
   c) Adjust capacity by incorporating maintenance and operations demands
   d) Adjust capacity by incorporating additional teams and personnel to boost productivity

39. You are planning along with your team. Which of the following would you expect on an agile project?
   a) Create initial scope, schedule, and cost range estimates that reflect detailed plans
   b) Create estimates with a high-level understanding of the effort necessary to deliver the project
   c) Develop a starting point for managing the project with as much minor detail as possible.
   d) Create a detailed Gant Chart and Budget for the project

40. As a result of being agile in one's approach, the likely effect will be that _______
   a) Scope and requirements are estimated just-in-time during sprint planning
   b) Scope and requirements are estimated just-in-time during the retrospective
   c) Scope is defined and redefined throughout the project
   d) Scope and requirements are estimated during sprint review with customers

41. You are a servant leader on an agile project working with other team members to refine the project scope, schedule and cost range estimates. What does this best describe?

a) Backlog Preparation
b) Backlog Planning
c) Story Mapping
d) Backlog Refinement

42. Your team has 3 user stories to accomplish.
A = 28 story points
B = 26 story points
C - 34 story points
Your team's velocity is 38 story points.
How many sprints would you need to complete all 3 stories?

a) 2.3
b) 4
c) 3
d) 3.5

43. You're working with an agile team on a flow-based agile project. You have been given the task of cost reduction and effectiveness. What should you do next to achieve this?

a) Sprint backlog
b) Value stream analysis
c) Burndown chart
d) Information radiator

44. You're working with an agile team on a flow based project. You have been given the task of cost reduction and effectiveness. What should you do next to achieve this?

a) Define the problem, identify the root cause, generate alternatives, choose the best alternative
b) Identify the root cause, define the problem, choose the best alternative, generate alternatives
c) Generate alternatives, define the problem, choose the best alternative, identify the root cause
d) Verify alternative worked, choose the best alternative, generate alternatives, identify the root cause

45. Your team is working on an agile approach with the following configuration:
Red, green, clean which agile approach is this most likely to be?

a) Behavior driven development
b) Test driven development
c) Wireframes
d) Personas

46. Your team is working on an agile approach with the following configuration:
"Given, When, Then"
Which agile approach is this most likely to be?

a) Writing acceptance tests
b) Running an automated unit test
c) Creating Wireframes
d) Persona Permutation Analysis

47. Your team has spent the last two weeks refactoring code that was written 3 iterations ago. Although the system worked, the operations team was finding difficulty servicing the code. What does this best describe in agile?

a) Defect
b) Cost of delay
c) Technical debt
d) Work in progress

48. Your team is using a test driven development approach. Which of the following orders of work would you expect?

a) Write test, write code, refactor code
b) write code, write test, refactor code
c) refactor test, write code, run test
d) write code, refactor code, run test

49. The potentially shippable increment for this brand was tested and it worked. However, when coupled with the increment from the previous Sprint, it did not work as expected. What could have been done to prevent this?

a) Continuous integration
b) continuous testing
c) continuous improvement
d) continuous coding

The best solution to this problem is: Continuous integration
Continuous Integration: A practice in which each team member's work products are frequently integrated and validated with one another.

50. Your team is in the middle of an iteration. The agile method being used is a flow based agile approach. You discover huge bottlenecks and swelling at key points in the workflow. What is the most likely cause for these problems?

a) Work in progress limits were set to low
b) Work in progress limits were set too high
c) The team is preoccupied with quality concerns
d) The product owner is incompetent

Domain 6 Task 2: Identify threats and issues by educating and engaging the team at various points in the project in order to resolve them at the appropriate time and improve processes that caused issues.

AGILE CERTIFIED PRACTITIONER HALF-MARATHON (ANSWERS)

1. Your senior executives are exhibiting behaviors that are in direct contrast to the agile manifesto. You have historically used a traditional approach and only recently started employing agile frameworks. What should you do as a servant leader/scrum master in the firm?

The correct answer is: Model agile principles and discuss with stakeholders

It is not the product owner's primary role to do this, however it is something the Scrum Master is expected to do.

Scrum Guide 2020: "The Scrum Master serves the organization in several ways, including: Leading, training, and coaching the organization in its Scrum adoption."

Why: Agile Value #1 and PMI-ACP Domain 1, Task 1

Advocate for agile principles by modeling those principles and discussing agile values in order to develop a shared mindset across the team as well as between the customer and the team.

2. Your senior executives are exhibiting behaviors that are in direct contrast to the agile manifesto. You have historically used a traditional approach and only recently started employing agile frameworks. What should you do as a servant leader/scrum master in the firm?

The correct answer is: Build common knowledge around agile practices and terminology being used in order to work effectively.

This is the role Scrum Masters and Servant Leaders should play. Also note that Kanban and Scrum are not the only Agile frameworks.

Why: Agile Value #1 and PMI-ACP Domain 1, Task 2

Help ensure that everyone has a common understanding of the
values and principles of agile and a common knowledge around the agile practices and terminology being used in order to work effectively.

3. Your senior executives are exhibiting behaviors that are in direct contrast to the agile manifesto. You have historically used a traditional approach and only recently started employing agile frameworks. As an agile leader, which of the following is your responsibility?

The correct answer is: All of the options.

All mentioned aspects are things an Agile leader should do.

Why: Agile Value #1 and PMI-ACP Domain 1, Task 3: Support change at the system or organization level by educating the organization and influencing processes, behaviors, and people in order to make the organization more effective and efficient.

4. Your agile team is struggling on the project. Transparency is a huge issue and the team has lost stakeholder trust. Estimates seem to be secrets and real progress cannot be seen by the stakeholders. What is the best course of action?

Answer: Use information radiators

Why: PMI-ACP Content Outline Domain 1 Task 4

Practice visualization by maintaining highly visible information radiators showing real progress and real team performance in order to enhance transparency and trust.

The other options do not remedy the transparency issue because they are all reactive (after work has occurred in acceptance, demos and retrospectives - where stakeholders would not likely be in attendance anyway.)

5. Having just converted to agile approaches, your team has made a number of mistakes using Scrum and is struggling with a sporadic demand pattern. What is the best thing to do?
The correct answer is: Flow-based agile
Why: The team is already struggling with an iterative Agile approach (Scrum), so that cannot be the best answer. It is ok to make mistakes so a world of "no mistakes" is an illusion in Agile. Agile thrives on learning through mistakes and errors to improve.
Agile Practice Guide Page 32: Many teams find that using a cadence (in the form of a regular timebox) helps them demo, retrospect, and take in new work. In addition, some teams need more flexibility in their acceptance of more work. Teams can use flow-based agile with a cadence to get the best of both worlds.

PMI-ACP Content Outline Domain 1 Task 5
Contribute to a safe and trustful team environment by allowing everyone to experiment and make mistakes so that each can learn and continuously improve the way he or she works.

6. Having just converted to agile approaches, your team is building a new product. The team may need feedback on a prototype to glean customer feedback to understand what they need to provide for subsequent delivery of the final finished feature . What is the best thing to do?
The correct answer is: Minimum viable product (the keyword here is prototype).
Why: Agile Practice Guide Page 23
Completeness and delivery are subjective. The team may need feedback on a prototype and may then choose to deliver a minimum viable product (MVP) to a subset of customers. The customers' feedback helps the team to learn what they need to provide for subsequent delivery of the final finished feature.
PMI-ACP Content Outline Domain 1, Task 6: Enhance creativity by experimenting with new techniques and process ideas in order to discover more efficient and effective ways of working.

7. Having just converted to agile approaches, your team is involved in the progressive elaboration of project requirements. They are involved in the ongoing activity in which the team collaboratively reviews, updates, and writes requests/requirements to satisfy the need of the customer request. Which action does this best describe?
The correct answer is: Backlog refinement
Why: Agile Practice Guide Page 150
In iteration-based agile, the product owner often works with the team to prepare some stories for the upcoming iteration during one or more sessions in the middle of the iteration. The purpose of these meetings is to refine enough stories so the team understands what the stories are and how large the stories are in relation to each other.
Domain 1 Task 7: Encourage team members to share knowledge by collaborating and working together in order to lower risks around knowledge silos and reduce bottlenecks.

8. Having just converted to agile approaches, your team is involved in the progressive elaboration of project requirements. As you proceed, certain individuals appear to naturally have more influence than others. What should you do as the servant leader?
The correct answer is: Encourage emergent leadership
Why: Emergent leadership refers to a situation where someone on the team is not appointed as a leader in a certain area over others, but that person steps up to a higher-level of accountability over time naturally through inter-team interactions. It empowers team members to make decisions outside a firm or business unit's traditional structure.

Domain 1 Task 8: Encourage emergent leadership within the team by establishing a safe and respectful environment in which new approaches can be tried in order to make improvements and foster self-organization and empowerment.

9. Having just converted to agile approaches, your team is involved in the progressive elaboration of project requirements. What should you do as the servant leader?
The correct answer is: Promote collaboration and conversation between teams
Why: Agile Practice Guide Page 35: Servant leaders promote collaboration and conversation within the team and between teams. For example, a servant leader helps to expose and communicate bottlenecks inside and between teams. Then the teams resolve those bottlenecks.
Domain 1 Task 9: Practice servant leadership by supporting and encouraging others in their endeavors so that they can perform at their highest level and continue to improve.

10. In agile, when ordering the backlog, it could be done primarily by:
The correct answer is: Value and benefit
Why: Backlogs are generally ordered primarily by value. All the other options work more as categorizations. MOSCOW and KANO enable categorization but you could still have 10 "must-haves" or 20 "delighters". Therefore, prioritization by elements (value, benefits, risks, cost of delay, NPV, ROI and other parameters) should also occur through a critical and decisive review of the backlog by the product owner with the team's help.

11. You are using an archetype user representing a set of similar end users described with their goals, motivations, and representative personal characteristics. What does this best describe?
The correct answer is: Personas
Why: Agile Practice Guide Page 153
Personas. An archetype user representing a set of similar end users described with their goals, motivations, and representative personal characteristics.

12. You are a servant leader on an agile project. Towards the end of the second iteration, you discover that a new stakeholder is in attendance at the Sprint review. This is the first time you are encountering this stakeholder. What course of action would most likely have prevented this surprise?
The correct answer is: Identifying and engaging stakeholders through periodic reviews
PMI-ACP Content Outline Domain 3 Task 1: Identify and engage effective and empowered business stakeholder(s) through periodic reviews in order to ensure that the team is knowledgeable about stakeholders' interests, needs, and expectations.
This covers the widest band of possibilities. The other options are too decisive or restrictive. Stakeholders can be identified at any time.

13. You work for a Fortune 500 firm in London. Over the past 15 years, you have been a program manager using predominantly predictive project management approaches. Recently, the CEO announced a re-organization and a new agile approach to planning, and you are now an Agile Certified Practitioner and Scrum Master.
What is the best way to identify and engage stakeholders?
The correct answer is: Promoting knowledge sharing early and through the project
Why: This is the most agile option and it is standard and proactive. The best way to engage stakeholders is not by directing them to view information.

PMI-ACP Content Outline Domain 3 Task 2: Identify and engage all stakeholders (current and future) by promoting knowledge sharing early and throughout the project to ensure the unimpeded flow of information and value throughout the lifespan of the project.

14. An agile transformation has just taken place in your firm after 4 weeks of training senior management (VPs, Directors and Managers). Recently, the CEO also announced an entire re-organization and this new agile approach. You begin using agile frameworks on your product delivery.
In your third iteration, you realize that stakeholders are not engaged. What should you do first?
The correct answer is: Apply servant leadership to work with this stakeholder (and possibly product owner).
Agile Practice Guide Page 59
PMI-ACP Content Outline Domain 3 Task 3: Establish stakeholder relationships by forming a working agreement among key stakeholders in order to promote participation and effective collaboration.

15. An Agile team discovered during the first couple of iterations that there was a gap in the expectations of the stakeholders and the deliverables produced by the team.
Which of the following would be most effective in reducing the gap?
The correct answer is: Continually assess changes in the organization to ensure stakeholder engagement.
The other options are either too prescriptive or not agile enough.
PMI-ACP Content Outline Domain 3 Task 4: Maintain proper stakeholder involvement by continually assessing changes in the project and organization in order to ensure that new stakeholders are appropriately engaged.

16. An Agile team needs services of a technical expert to run certain tasks. Some of the team has concerns about this. They feel this arrangement might impede their progress and raised this concern to the Team Facilitator. What is the best course of action for the Team Facilitator?
The correct answer is: Use Highsmith decision spectrum or voting for collaborative team decision making full team involvement and decision-making collaboration is the best option.
PMI-ACP Content Outline Domain 3 Task 5: Establish collaborative behaviors among the members of the organization by fostering group decision making and conflict resolution in order to improve decision quality and reduce the time required to make decisions.

17. As a facilitator of an agile team, you are seeking to establish a shared vision of the various project increments with the team and senior management. How should you go about this task?
The correct answer is: Develop an agile project charter to share the vision with stakeholders.
All the other options contain unnecessary, ill-timed or bogus steps.
PMI-ACP Content Outline Domain 3 Task 6: Establish a shared vision of the various project increments (products, deliverables, releases, iterations) by developing a high level vision and supporting objectives in order to align stakeholders' expectations and build trust.

18. While working together with your agile team, team members share concerns about a lack of trust among stakeholders. There seems to be an underlying belief that the team will fail. What should you do as a servant leader?
The correct answer is: Create an environment with a shared understanding of deliverables, success criteria and acceptable trade-offs.
Why: A daily review of acceptance criteria is unnecessary as this should have been done prior to the current sprint. Communicating mainly at sprint review is reactive. Encouraging the team to work over-time is against agile culture. Remember, per the Agile Manifesto, they should be able to maintain a constant pace indefinitely!

PMI-ACP Content Outline Domain 3 Task 7: Establish and maintain a shared understanding of success criteria, deliverables, and acceptable trade-offs by facilitating awareness among stakeholders in order to align expectations and build trust.

19. Based on the Agile PMO's advice, your agile team of 45 members, has begun to scale practices being used in the firm. As they do this, they receive a lot of stakeholder pushback. Leadership of the technology group has accused the team of wasting valuable development time in too many scrum-of-scrums. What is the best thing for you to do as the servant leader?
The correct answer is: Mentor stakeholders and be transparent regarding the work process and status by communicating elements of concern, impediments, quality and risks.
Why: Avoiding ALL hidden risks is generally not possible. The other options have incorrect or non-agile suggestions.
PMI-ACP Content Outline Domain 3 Task 8: Provide transparency regarding work status by communicating team progress, work quality, impediments, and risks in order to help the primary stakeholders make informed decisions.

20. Your production team has moved from a predictive method to an agile approach. You are using the concepts of work in progress, cost of delay, and cycle time to provide a high-quality product. As the project proceeds, stakeholders ask for status. What should you do?
The correct answer is: Give stakeholders forecasts at a level of detail that balances benefits of certainty and adaptability so they can effectively plan.
Providing forecasts at a granular level of detail, planning the entire project, negating certainty, and sticking to original plans are all against the agile mindset. Balancing certainty and adaptability is key.
PMI-ACP Content Outline Domain 3 Task 9: Provide forecasts at a level of detail that balances the need for certainty and the benefits of adaptability in order to allow stakeholders to plan effectively.

21. A few weeks ago, you were working on the social contract foster team coherence and strengthen team members' commitment to shared outcomes. A few weeks later, your team is in a heightened conflict stage where you need to refer to this contract. Which stage of team development are they likely in?
The correct answer is: Storming
The team is most likely in Storming where conflict is most likely to occur.
PMI-ACP ECO Domain 4 Task 1: Cooperate with the other team members to devise ground rules and internal processes in order to foster team coherence and strengthen team members' commitment to shared outcomes.

22. Your team has found themselves in hyper-norming. Everyone agrees about everything even when it is wrong. Which of the following should the servant leader uphold?
The correct answer is: Constructive disagreement. Hyper-conflict is a made-up term in this context.
PMI-ACP ECO: Domain 4 Task 1: Cooperate with the other team members to devise ground rules and internal processes in order to foster team coherence and strengthen team members' commitment to shared outcomes.

23. As a servant leader you are helping to create a team of generalizing specialists that has interpersonal and technical skills needed to achieve the project objectives. What is the main purpose and objective for these actions?
The correct answer is: To create business value with minimal delay. Some of the other options repeat what the question refers to as "generalizing specialists."

PMI-ACP ECO: Domain 4 Task 2: Help create a team that has the interpersonal and technical skills needed to achieve all known project objectives in order to create business value with minimal delay. Also see PMI-ACP ECO: Domain 7 Task 4

24. Your agile team is performing at its peak. All team members are fully engaged and involved. Team members need little direction and support. Which leadership approach would best suit the team at this stage?

The correct answer is: Delegating

This conforms to the Hersey-Blanchard model.

In order to make an effective cycle, a leader needs to motivate followers properly by adjusting their leadership style to the development level of the person. Blanchard postulates that Enthusiastic Beginners (D1) need a directing leadership style while Disillusioned Learners (D2) require a coaching style. He suggests that Capable but Cautious Performers (D3) respond best to a Supporting leadership style and Self-reliant Achievers need leaders who offer a delegating style. http://en.wikipedia.org/wiki/Situational_leadership_theory

PMI-ACP ECO: Domain 4 Task 4: Contribute to self-organizing the work by empowering others and encouraging emerging leadership in order to produce effective solutions and manage complexity.

25. Your agile team is very fragile and performing poorly. All team members are disengaged and worried. Team members need a lot of direction and support. Which leadership approach would best suit the team at this stage?

The correct answer is: Coaching

This conforms to the Hersey-Blanchard model.

In order to make an effective cycle, a leader needs to motivate followers properly by adjusting their leadership style to the development level of the person. Blanchard postulates that Enthusiastic Beginners (D1) need a directing leadership style while Disillusioned Learners (D2) require a coaching style. He suggests that Capable but Cautious Performers (D3) respond best to a Supporting leadership style and Self-reliant Achievers need leaders who offer a delegating style. http://en.wikipedia.org/wiki/Situational_leadership_theory

PMI-ACP ECO: Domain 4 Task 4: Contribute to self-organizing the work by empowering others and encouraging emerging leadership in order to produce effective solutions and manage complexity.

26. As the servant leader of an agile team, you observe that a team member has the need for very little direction but a lot of support. What is the best approach for leading in this circumstance?

The correct answer is: Participating. This is an alternate term for Supporting. When in doubt, use process of elimination! The titles for three of these styles differ depending on which version of the model is used. Delegating (Same across models), Participating (or Supporting), Selling (Coaching) and Telling (Directing).

PMI-ACP ECO: Domain 4 Task 4: Contribute to self-organizing the work by empowering others and encouraging emerging leadership in order to produce effective solutions and manage complexity.

27. Your project team complains to you about certain stakeholders who have breached the team space, making requests for work not included in the sprint. As a self-organizing team, what is your best course of action as a servant leader?

The correct answer is: Shield the team from interruptions. This is the job of the Scrum Master or Servant Leader.

The other options are either not in the spirit of Agile and customer collaboration.

PMI-ACP ECO: Domain 4 Task 7

28. Your project team complains to you about certain project and release aspects. As a self-organizing team, what is your best course of action as a servant leader?

The correct answer is B: Work with the product owner and stakeholders to find resolve. This is the best option since the product owner is responsible for the release aspects.

PMI-ACP ECO: Domain 4 Task 4: Contribute to self-organizing the work by empowering others and encouraging emerging leadership in order to produce effective solutions and manage complexity.

29. You are a servant leader on an agile team deep in the middle of development. Which of the following would your role most likely be?

The correct answer is: Carrying food and water

This is what a servant leader could do to serve. The other listed items are things the team should do.

Also conflict issues should not be solved by anyone other than those involved.

PMI-ACP ECO: Domain 4 Task 5: Continuously discover team and personal motivators and demotivators in order to ensure that team morale is high and team members are motivated and productive throughout the project.

30. Your agile team has put in a lot of work towards a product release. Which of the following best describes an immediate reward system that you would employ for the team at this time?

The correct answer is: Treating the team to dinner at a five-star resort

The key word is "for the team at this time".

Rewarding the team collectively is the best option in Agile environments.

PMI-ACP ECO Domain 4 Task 5: Continuously discover team and personal motivators and demotivators in order to ensure that team morale is high and team members are motivated and productive throughout the project.

31. Your team was in the same location previously, but now they are virtually co-located. Which of the following would you expect as a result of this colocation?

The correct answer is: Greater challenges in conflict resolution

They are virtually co-located which would make conflict resolution more challenging.

PMI-ACP ECO: D4 Task 6: Facilitate close communication within the team and with appropriate external stakeholders through co-location or the use of collaboration tools in order to reduce miscommunication and rework.

32. While working together with your agile team, team members share concerns about stakeholder's perception of commitment and goal alignment. What should you do as a servant leader?

The correct answer is: Share the project vision and ensure the team understands how their objectives fit the overall goals

The other options are either incorrect things to do faced with this problem or not in the spirit of Agile.

PMI-ACP Content Outline Domain 4 Task 8: Participate in aligning project and team goals by sharing project vision in order to ensure the team understands how their objectives fit into the overall goals of the project.

33. Which of the following is true about Agile teams?

The correct answer is: When people are unavailable, the product owner understands that the team has reduced capacity.

Time away from work such as holidays, time off and emergencies reduces the team's capacity. As a result, the passage of time could be much longer.

ACP ECO: Domain 4, Task 9: Encourage the team to measure its velocity by tracking and measuring actual performance in previous iterations or releases in order for members to gain a better understanding of their capacity and create more accurate forecasts.

34. As a servant leader, it is important that you employ emotional intelligence. Which of the following best describes emotional intelligence?
The correct answer is: Being able to control one's emotions and influence the emotions of other individuals and groups.
This is the best answer.
ACP Toolkit (ACP ECO Page 14)

35. As a team using an agile framework, which of the following approaches to planning would you expect?
The correct answer is: Plan at multiple levels (strategic, release, iteration, daily). This is the most inclusive option because it contains strategic and daily planning.
PMI-ACP Content Outline Domain 5 Task 1
Remember the concept of the Agile Planning Onion

36. You work for a Fortune 500 firm in London. Recently, the CEO announced a re-organization and a new agile approach to planning, and you are now an Agile Certified Practitioner. What is the best way to make planning activities visible and transparent?
The correct answer is: Encourage participation of key stakeholders and publishing planning results
Why: This is the best option Because the other options are either wrong or prescriptive. Stakeholders do not participate in creating burndown charts or by planning tasks that the team undertakes.
PMI-ACP Content Outline Domain 5 Task 2

37. Your team is practicing scrum as an agile framework. Which scrum event in your sprint, will give your team the greatest opportunity to inspect and adapt the project plan to reflect changes in requirements, schedule, budget, and shifting priorities?
The correct answer is: Retrospective
This is the best option because it happens at the end of the Sprint when everything else including the Sprint review has been done. This gives the greatest opportunity to inspect and adapt all the events and variables mentioned.
APG, Page 50: The single most important practice is the retrospective because it allows the team to learn about, improve, and adapt its process. Retrospectives help the team learn from its previous work on the product and its process.
ACP ECO: Domain 5 Task 4
Adapt the cadence and the planning process based on results of periodic retrospectives about characteristics and/or the size/complexity/criticality of the project deliverables in order to maximize the value.

38. As you consider range estimates on your agile project, which of the following ideas should you take into account?
The correct answer is: Adjust capacity by incorporating maintenance and operations demands
ACP ECO: Domain 5 Task 7
Adjust capacity by incorporating maintenance and operations demands and other factors in order to create or update the range estimate.

39. You are planning along with your team. Which of the following would you expect on an agile project?
The correct answer is: Create estimates with a high-level understanding of the effort necessary to deliver the project
All the other statements are incorrect. You would not start an agile project with as much minor detail as possible. Detailed plans are not used in agile.
ACP ECO: Domain 5 Task 8
Create initial scope, schedule, and cost range estimates that reflect current high level understanding of the effort necessary to deliver the project in order to develop a starting point for managing the project.

40. As a result of being agile in one's approach, the likely effect will be that _______
The correct answer is: Scope is defined and redefined throughout the project
Agile Practice Guide Page 91, All the other statements are incorrect.
The key point to note is this does not happen only during sprint planning or a retrospective. It is constantly redefined.
ACP ECO: Domain 5 Task 9
Refine scope, schedule, and cost range estimates that reflect the latest understanding of the effort necessary to deliver the project in order to manage the project.

41. You are a servant leader on an agile project working with other team members to refine the project scope, schedule and cost range estimates. What does this best describe?
The correct answer is: Backlog Refinement
ACP ECO: Domain 5 Task 9
Refine scope, schedule, and cost range estimates that reflect the latest understanding of the effort necessary to deliver the project in order to manage the project.
APG P. 150: Backlog Refinement. The progressive elaboration of project requirements and/or the ongoing activity in which the team collaboratively reviews, updates, and writes requirements to satisfy the need of the customer request.

42. Your team has 3 user stories to accomplish.
A = 28 story points
B = 26 story points
C - 34 story points
Your team's velocity is 38 story points.
How many sprints would you need to complete all 3 stories?
The correct answer is: 88/38 = 2.3
You will need at least 3 sprints to get the work done.
0.3 would count as one additional sprint Because we do not count sprints in fractions. It is all or nothing because we would not stop a Sprint a portion of the way. ACP ECO: Domain 5 T10

43. You're working with an agile team on a flow-based agile project. You have been given the task of cost reduction and effectiveness. What should you do next to achieve this?
The best solution to this problem is: Value Stream Analysis.
Also called: value stream analysis, lean process mapping, value stream mapping (VSM) is defined as a lean tool that employs a flowchart documenting every step in the process.
Many lean practitioners see VSM as a fundamental tool to identify waste, reduce process cycle times, and implement process improvement.

44. You're working with an agile team on a flow based project. You have been given the task of cost reduction and effectiveness. What should you do next to achieve this?
The best solution to this problem is: Define the problem, identify the root cause, generate alternatives, choose the best alternative
This is simple logic. Problems must be defined first before identifying their root cause. When trying to solve a problem in a retrospective , there must first be a defined problem before identifying root causes, and generating alternatives.

45. Your team is working on an agile approach with the following configuration:
Red, green, clean which agile approach is this most likely to be?
The best solution to this problem is: Test driven development
Red, green, clean is indeed the format for test driven development.
In TDD, each test case fails initially: This ensures that the test really works and can catch an error. Once this is shown, the underlying functionality can be implemented. This has led to the "test-driven development mantra", which is "red/green/refactor", where red means fail and green means pass. Test-driven development constantly repeats the steps of adding test cases that fail, passing them, and refactoring.
https://en.wikipedia.org/wiki/Test-driven_development

46. Your team is working on an agile approach with the following configuration:
"Given, When, Then"
Which agile approach is this most likely to be?
The best solution to this problem is: Writing acceptance tests
The Given-When-Then formula is a template intended to guide the writing of acceptance tests for a User Story:
(Given) some context
(When) some action is carried out
(Then) a particular set of observable consequences should obtain
An example:

- *Given my bank account is in credit, and I made no withdrawals recently,*
- *When I attempt to withdraw an amount less than my card's limit,*
- *Then the withdrawal should complete without errors or warnings*

47. Your team has spent the last two weeks refactoring code that was written 3 iterations ago. Although the system worked, the operations team was finding difficulty servicing the code. What does this best describe in agile?
The best solution to this problem is: Technical Debt
APG Page 154, Technical Debt. The deferred cost of work not done at an earlier point in the product life cycle.

48. Your team is using a test driven development approach. Which of the following orders of work would you expect?
The best solution to this problem is: Write test, write code, refactor code
Test-Driven Development: A technique where tests are defined before work is begun, so that work in progress is validated continuously, enabling work with a zero-defect mindset. The order of work is write test, write code, and refactor code.

49. The potentially shippable increment for this brand was tested and it worked. However, when coupled with the increment from the previous Sprint, it did not work as expected. What could have been done to prevent this?

The best solution to this problem is: Continuous integration

Continuous Integration: A practice in which each team member's work products are frequently integrated and validated with one another.

50. Your team is in the middle of an iteration. The agile method being used is a flow based agile approach. You discover huge bottlenecks and swelling at key points in the workflow. What is the most likely cause for these problems?

The best solution to this problem is: Work in progress limits were set too high.

A Kanban board helps the team to further improve its effectiveness by visualizing the flow of work, making impediments easily visible, and allowing flow to be managed by adjusting work in process limits. Having too high WIP limits means that your team is probably working on multiple tasks, switching context all the time, and not meeting the deadlines.

Glossary of Common Agile Terms

1. Acceptance Test-Driven Development (ATDD) - Test criteria created before delivery.
2. Agile - Values and principles for agile practices.
3. Agile Coach - Mentors teams through agile transformation.
4. Agile Life Cycle - Incremental approach to project delivery with frequent releases.
5. Agile Manifesto - Official definition of agile values and principles.
6. Agile Mindset - Way of thinking based on Agile Manifesto.
7. Agile Practitioner (Agilist) - Embraces agile values and collaborates.
8. Agile Principles - Twelve principles of agile project delivery.
9. Agile Unified Process - Simplified approach to developing software using agile techniques.
10. Anti-Pattern - Flawed pattern of work to be avoided.
11. Automated Code Quality Analysis - Testing for bugs and vulnerabilities.
12. Backlog (Product Backlog) - Collaborative list of project requirements.
13. Backlog Refinement - Elaborating project requirements to satisfy customers.
14. Behavior-Driven Development (BDD) - Validation practice using test-first principles.
15. Blended Agile - Combining two or more agile frameworks.
16. Blocker (Impediment) - Hindrance or obstacle to team progress.
17. Broken Comb (Paint Drip) - Person with varying skill sets.
18. Burndown Chart - Graphical representation of remaining work in a timebox.
19. Burnup Chart - Graphical representation of completed work toward release.
20. Business Requirement Documents (BRD) - Listing of project requirements.
21. Cadence - Rhythm of execution, also known as a timebox.
22. Collective Code Ownership - Any team member can modify work products.
23. Continuous Delivery - Delivering feature increments immediately to customers.
24. Continuous Integration - Frequently integrating and validating team work products.
25. Cross-Functional Team - Team with all skills for valuable product increments.
26. Crystal Family of Methodologies - Agile methods for adaptability.
27. Daily Scrum (Daily Standup) - Brief daily collaboration meeting.
28. Definition of Done (DoD) - Checklist for customer-ready deliverables.
29. Definition of Ready (DoR) - Checklist for user-centric requirements.
30. DevOps - Practices improving development and operations staff collaboration.
31. Disciplined Agile (DA) - Incremental and iterative solution delivery.
32. Dynamic Systems Development Method (DSDM) - Agile project delivery framework.
33. eXtreme Programming (XP) - Agile software development method.
34. Feature-Driven Development (FDD) - Agile method driven by feature value.
35. Fit for Purpose - Product suitable for its intended purpose.
36. Fit for Use - Product currently usable for its intended purpose.
37. Impediment (Blocker) - Obstacle to team objectives.
38. Increment - Functional, tested, and accepted deliverable.

39. Incremental Life Cycle - Approach providing immediate use deliverables.
40. Information Radiator - Visible display for organization knowledge sharing.
41. I-shaped - Person with single area of specialization.
42. Iteration - Timeboxed cycle of development for product deliverables.
43. Iterative Life Cycle - Approach allowing feedback for unfinished work.
44. Kaizen Events - Collaborative events to improve the system.
45. Kanban Board - Visualization tool to improve workflow efficiency.
46. Kanban Method - Agile method for knowledge work using visual management.
47. Large Scale Scrum (LeSS) - Scrum framework for large product development.
48. Lean Software Development (LSD) - Applying lean principles to software development.
49. Life Cycle - Process of imagining, creating, and using a product.
50. Mobbing - Simultaneous focus by multiple team members on work.
51. Organizational Change Management - Structured approach to transition organizations.
52. Paint-Drip (Broken Comb) - Person with varied skills across specializations.
53. Pair Work (Pair Programming) - Two team members work simultaneously on tasks.
54. Personas - Archetype user with goals, motivations, and characteristics.
55. Pivot - Planned course correction to test new hypotheses.
56. Plan-Do-Check-Act (PDCA) - Iterative method for process control and improvement.
57. Plan-Driven Approach (Predictive Approach) - Traditional work management approach.
58. Predictive Life Cycle - Traditional sequential approach to work management.
59. Project Management Office (PMO) - Management structure for project governance.
60. Product Backlog - List of user-centric requirements for a product.
61. Product Owner (Service Request Manager) - Person responsible for maximizing product value.
62. Progressive Elaboration - Iterative process of increasing plan detail.
63. Release Planning - Collaborative planning of a product release.
64. Retrospective Prime Directive - Guiding principle for retrospective meetings.
65. Scaled Professional Scrum (SPS) - Framework for Scrum in complex environments.
66. Scrum of Scrums - Technique for scaling Scrum to multiple teams.
67. Scrum Team - Combination of team roles in Scrum.
68. Self-Organizing Team - Cross-functional team with fluid leadership.
69. Servant Leadership - Leading through service to team members.
70. Service Request Manager - Person responsible for ordering service requests.
71. Siloed Organization - Organization structured to contribute a subset of required aspects.
72. Smoke Testing - Lightweight testing for important system functions.
73. Specification by Example (SBE) - Collaborative approach to defining requirements.
74. Spike - Short time interval for research or prototyping.
75. Sprint - Timeboxed iteration in Scrum.
76. Sprint Backlog - List of work items for a Scrum sprint.
77. Sprint Planning - Collaborative planning for a Scrum sprint.

78. Story Point - Unit-less measure for relative estimation.
79. Swarming - Collective focus on resolving a specific impediment.
80. Technical Debt - Deferred cost of unfinished work when cutting corners.
81. Test-Driven Development - Defining tests before work for continuous validation.
82. Timebox - Fixed period of time for work.
83. T-shaped - Person with deep specialization and broad abilities.
84. User Story - Brief description of deliverable value for a specific user.
85. User Story Mapping - Visual practice for organizing high-value features.
86. UX Design - Improving product usability and accessibility for better user experience.
87. Value Stream - Organizational construct focused on delivering value to customers.
88. Value Stream Mapping - Lean technique to analyze and improve product/service flow.

Made in United States
Orlando, FL
04 November 2026

90200486R00189